survival 2001
scenario from the future

survival 2001
scenario from the future

Henry E. Voegeli

John J. Tarrant

VNR VAN NOSTRAND REINHOLD COMPANY
NEW YORK CINCINNATI TORONTO LONDON MELBOURNE

Van Nostrand Reinhold Company Regional Offices:
New York Cincinnati Chicago Millbrae Dallas

Van Nostrand Reinhold Company International Offices:
London Toronto Melbourne

Library of Congress Catalog Card Number: 74-29310
ISBN: 0-442-28406-3

Manufactured in the United States of America

Published by Van Nostrand Reinhold Company
450 West 33rd Street, New York, N.Y. 10001

Published simultaneously in Canada by Van Nostrand Reinhold Ltd.
15 14 13 12 11 10 9 8 7 6 5 4 3 2 1

Library of Congress Cataloging in Publication Data

Voegeli, Henry Edward.
    Survival 2001.

    1. Power resources.   2. Power (Mechanics)
3. Energy conservation.   I. Tarrant, John J., joint
author.   II. Title.
TJ153.V58   621   74-29310
ISBN 0-442-28406-3

Contents

survival 2001
scenario from the future

Introduction

By the late 1970s it seemed to many people that the earth was dying; or at least the part that had been most favored by fortune—the United States of America.

As we look back from the vantage point of the first decade of the twenty-first century, it is not easy for us to realize this. It is particularly difficult for young people. They have read and heard the reminiscences; but these seem almost as wildly exaggerated as the tales of senseless wars like the 30-year Indochina conflict, the "limited" struggles along the USSR-China frontier in the 1980s, or the "lightning incursion" that took place in central Europe in 1992.

It is hard for us to understand the way it was then—the way the people thought and acted, the motivations that gave rise to actions that now seem incredible to us. Nevertheless, it is true that many people thought that the earth was dying then, choking in its own affluence and its own effluents. And there was considerable basis for these fears. What was then called the "ecology"—a term describing the relationship among the water, air, soil, flora, and fauna—was turning to poison.

This book constitutes an effort to help the student of today to understand what was happening then—and what was done to reverse the damaging processes. The book covers a number of the major concepts with which we are familiar now, concepts and programs affecting the air, the earth, and the water. It goes back as far as the 1970s—using the actual words and drawings of thinkers of that time—to show the magnitude of the problems confronting the people of the earth and how they began to work toward solutions.

The viable existence of the human race seems assured in the twenty-first century. This book has been produced as a reminder of how narrowly that victory was won, and how close the earth came to disaster.

The Houses We Live In

Whether it is hot or cold, we are quite comfortable indoors. This comfort is achieved without great cost in depletion of resources or in pollution of the environment.

Building practices are sensible nowadays. But there are people still living who remember quite a different situation. Buildings were impractical and wasteful—often dreadfully so. Although many people did not seem to care, some did. Here is the warning, from many decades ago, issued by one of those who cared:

Conservation can have a significant meaning in the selecting and operating of a home. It is quite reasonable to assume that future generations will make use of the cold in winter and the heat of summer. It is also probable that hills and mountains will be hollowed out for commerical and living space, thus taking advantage of the year-round temperature of 50 to 60° F.

Also of some intrinsic worth is the pride of home ownership. A well-designed, attractive house in a proper setting costs no more than one that is mis-shappen; and it has its reward in added resale value. The design should have recognizable characteristics of a style that has withstood the test of time, and one that is the most suitable for the site. For example, the Georgean or New England Colonial belongs on a village green, whereas the Chalet is natural to a forested area, overlooking a lake or stream.

Conservation of one's limited resources also demands seeking out economies in construction. Prefabrication of modular units holds out hope for cheaper, better-designed building elements for both the exterior and the interior. Useful plastics and other new materials are in the offing. Air quilting and pneumatics will probably come into practical use.

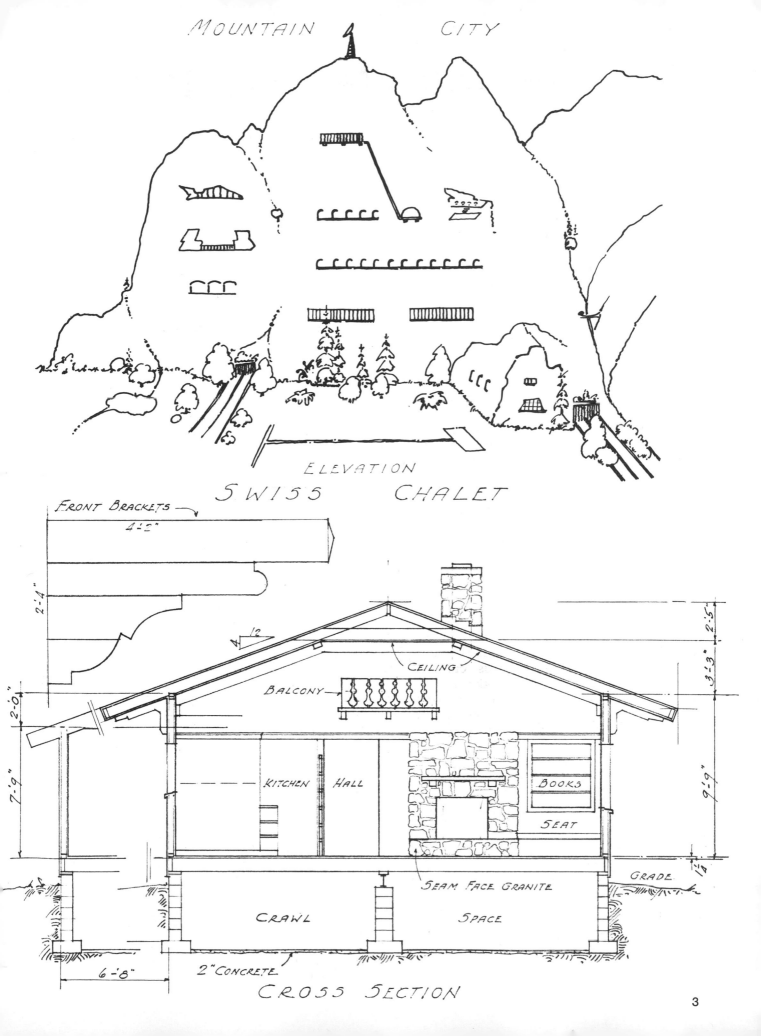

MOUNTAIN   CITY

ELEVATION

SWISS   CHALET

FRONT BRACKETS

4'-2"

2'-4"

2'-0"

7'-9"

6'-8"

12
4

CEILING

BALCONY

KITCHEN   HALL

BOOKS

SEAT

SEAM FACE GRANITE

CRAWL   SPACE

2" CONCRETE

GRADE

2'-5"
3'-3"
9'-9"

CROSS SECTION

3

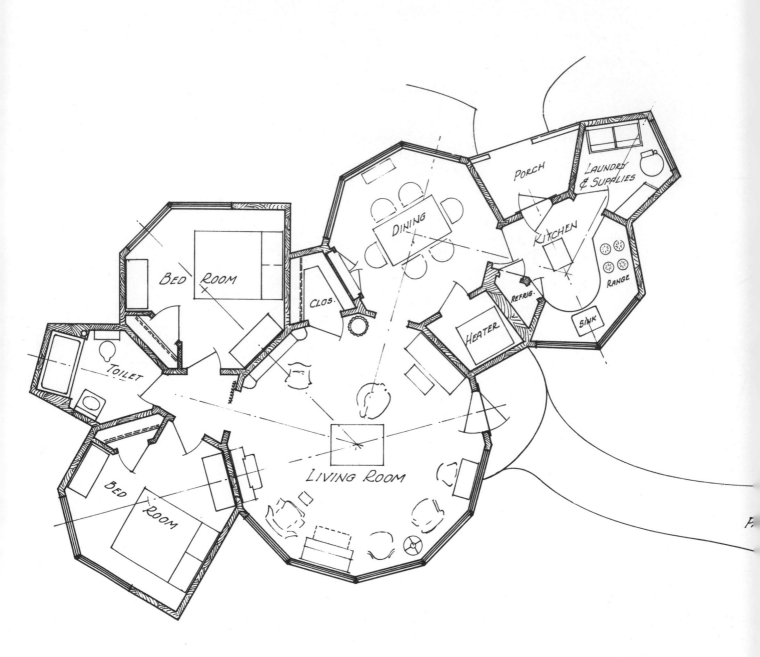

BED ROOM

DINING

PORCH

LAUNDRY & SUPPLIES

KITCHEN

CLOS.

REFRIG.

RANGE

TOILET

HEATER

SINK

LIVING ROOM

BED ROOM

- CELLULAR HOUSE -

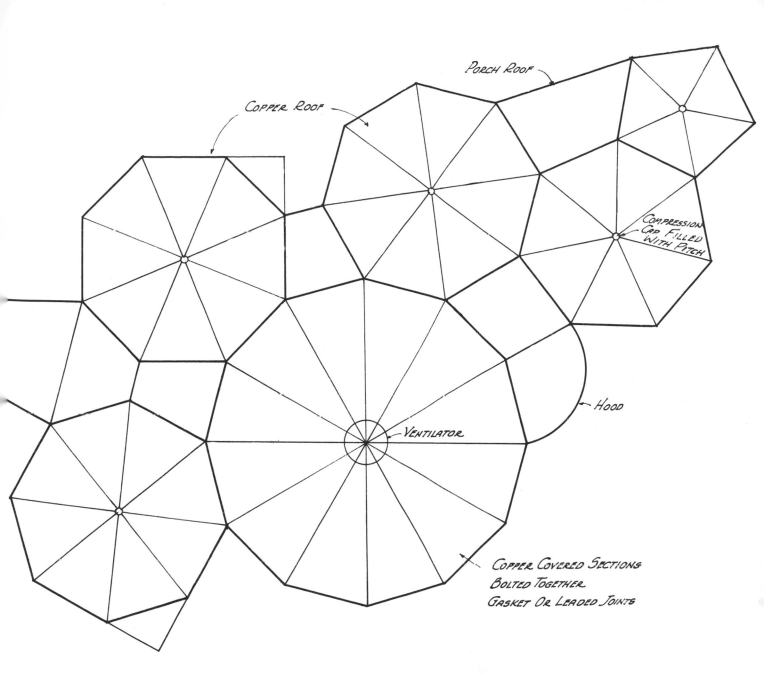

COPPER ROOF

PORCH ROOF

COMPRESSION CAP FILLED WITH PITCH

HOOD

VENTILATOR

COPPER COVERED SECTIONS
BOLTED TOGETHER
GASKET OR LEADED JOINTS

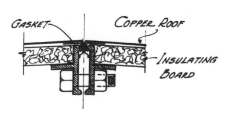

GASKET

COPPER ROOF

INSULATING BOARD

– METAL JOINT –

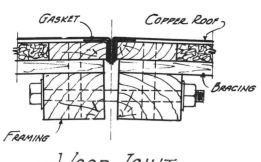

GASKET

COPPER ROOF

BRACING

FRAMING

– WOOD JOINT –

5

AIR CONDITIONING WITH THE HEAT PUMP

One of the ways of saving fuel in heating and cooling a house requires a heat pump. The heat pump is a machine that causes certain sensitive fluids to flip-flop from the liquid to the gaseous state, and back again by the manipulation of temperatures and pressures. That is the principle embodied in the ordinary household refrigerator. Like a refrigerator, air conditioning requires that heat be moved from one place to another; it requires a heat source and a heat sink. The earth, and a refrigerating compressor with a temperature range from about 25° F on the low side to around 145° F on the high side, are quite satisfactory.

About 50 feet below the surface of the earth the temperature is more or less constant, ranging from 50 to 60° F along the Atlantic Coast. This is a suitable heat source and heat sink for air conditioning. With a coefficient of performance (COP) of 3 (the compressor will yield three times the heat energy of the power consumed), the heat pump can compete with other methods of air conditioning, and at the same time conserve a great deal of fuel. (Nevertheless, air conditioning will become increasingly expensive and should be accompanied by a thorough and effective job of insulation.)

The operation of the heating cycle in air conditioning begins by starting the left water circulating pump, (the right pump is for the cooling cycle). The compressor is then put into service. Its function is to add the heat of compression to the sensible heat that is collected by the water in the deep underground piping, and which is then delivered to the convectors for heating the house, at a temperature of around 145°F. The water then passes through the second heat exchanger, the evaporator, where the remaining warmth is absorbed and delivered to the first heat exchanger, (the condenser). The removal of the heat lowers the temperature to 30°, and the chilled water sent through the piping in the earth to pick up more "black heat" on its way back to the compressor to repeat the cycle.

The cooling cycle begins at the evaporator heat exchanger, with water at 30° circulated through tube and fin, or cast-iron convectors to the house, and then passed through the condenser heat exchanger where again the collected heat is added and sent though the underground piping for stauration into the deep earth.

AIR CONDITIONING WITH THE HEAT PUMP

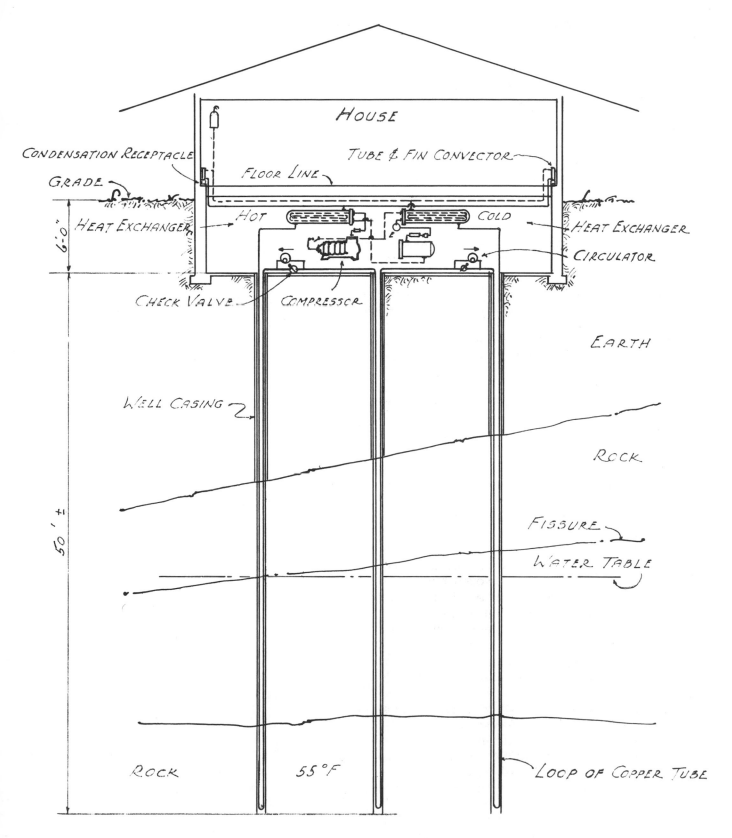

SCHEMATIC SECTION

CONSERVATION THROUGH INSULATION

In hundreds of thousands of our modern homes, office buildings, and factories, people live and work comfortably when the outside temperature approaches 0°F—and they do this without the benefit of much, if anything, in the way of heating apparatus. New buildings, large and small, are now designed so that they include only small auxiliary heaters and fans—to be used, for example, to warm up the building after it has been unoccupied for a time; say, over a weekend.

We are used to this. Therefore it is all the more interesting for us to contemplate the fact that buildings of a few decades ago—right up to the 1990s—all reguired heating plants that would seem enormously elaborate and cumbersome to us today. Fifty years ago in a city like Chicago, a large office building would have a huge central furnace system throbbing away throughout the cold months, pumping warmth to the spaces throughout the building. Much of this warmth was wasted—as was the warmth produced by the furnaces of private homes. Today the large heating plant is unnecessary.

The same is true of the large and complex cooling machinery that was—not so long ago—necessary to make buildings comfortable in warm weather.

In our enclosed spaces we remain comfortable in winter—*warmed by the heat from our own bodies*. And in summer we stay cool because the cooled air surrounding us remains that way.

The difference has been made by a revolutionary approach to a most mundane topic—*insulation*. For hundreds of years buildings had been insulated in more or less the same way. Here is how a contemporary observer describes the picture:

Insulating materials in use today commonly depend on minute dead-air spaces separated by material, often fibrous, having low thermal conductance. These materials range from fiberboard and bulk fibrous material to cork, and a one-inch thick piece of them has an insulating value approximately three times greater that of a one-inch air space, a single thickness of plate glass, or a three-quarters inch thick piece of wood. The coefficient of heat transmission for each is approximately .33 BTU/ft²-hr per °F difference of inside and outside termperature (disregarding air motion and moisture content of the insulation).

With a thermal conductivity per inch (the "k" factor) of 12 for stone and concrete, 8 for brick and terra-cotta, 1.13 for window glass, and 300 to 2600 for metals, conservation in heating and cooling is unthinkable.

BUILDING WALLS FOR COLD CLIMATES

CONVENTIONAL WALL

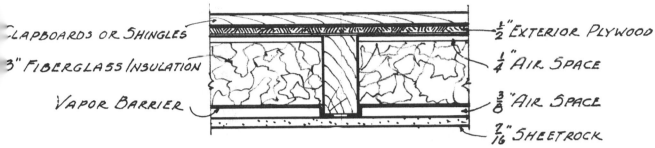

CLAPBOARDS OR SHINGLES

3" FIBERGLASS INSULATION

VAPOR BARRIER

$\frac{1}{2}$" EXTERIOR PLYWOOD

$\frac{1}{4}$" AIR SPACE

$\frac{3}{8}$" AIR SPACE

$\frac{7}{16}$" SHEETROCK

WALL WITH REFLECTIVE INSULATION

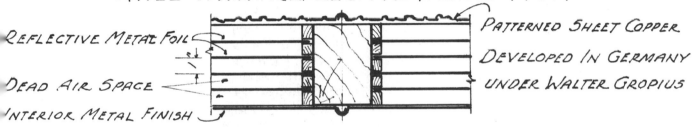

REFLECTIVE METAL FOIL

DEAD AIR SPACE

INTERIOR METAL FINISH

PATTERNED SHEET COPPER

DEVELOPED IN GERMANY

UNDER WALTER GROPIUS

AIR SPACES AND INSULATING BOARD

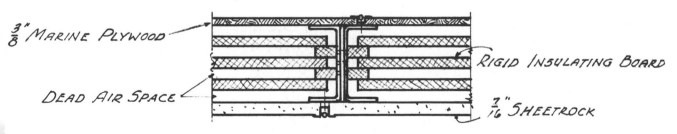

$\frac{3}{8}$" MARINE PLYWOOD

DEAD AIR SPACE

RIGID INSULATING BOARD

$\frac{7}{16}$" SHEETROCK

DOUBLE GLAZING

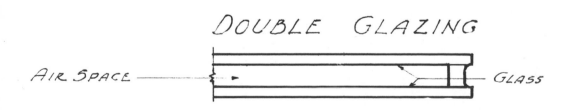

AIR SPACE

GLASS

As fuel became scarce and cost of heating oil soared, attention was directed toward the necessity for stopping the wasteful loss of desired warmth and coolness. The observer goes on:

A home should be at least as efficient as the present household refrigerator. A building in a hot climate should be heat reflective, have a minimum of heat storage or thermal capacity, and be dehumidified and appropriately air conditioned. Buildings in cold climates may have an advantage in their considerable mass, but only with ample, bone-dry insulation.

Since the body is a heat machine, it is theoretically possible to insulate an enclosure so that no heating system will be required. According to the drawing, an 8½ inch thickness of fiberglass insulation would slow down the heat flow so that the body heat from a proportionate 2½ adults in a 1000 cubic foot space would obviate the need of a heating system. Of course, the 1 to 1½ air changes per hour would have to be brought to room temperature by the heat of lamps, cooking, and other heat producing means.

In the new world of plastics, a sheet material that is impervious to water and air, and can be sweated and sealed by chemical and electronic means, offers an opportunity to incorporate the benefits of a vacuum or partial vacuum for insulating buildings.

A light bulb can hold a vacuum because of a structural form that prevents collapse. Likewise, a structural building panel can be made to sustain a vacuum by surrounding and enveloping it with a sheath of impervious material, having seams that are equally vacuum tight. An over-all conductance or "U" value as shown may not be attainable, but 1/5 that amount would be five times better than what we have today.

NO HEATING SYSTEM — 0° TO 70°

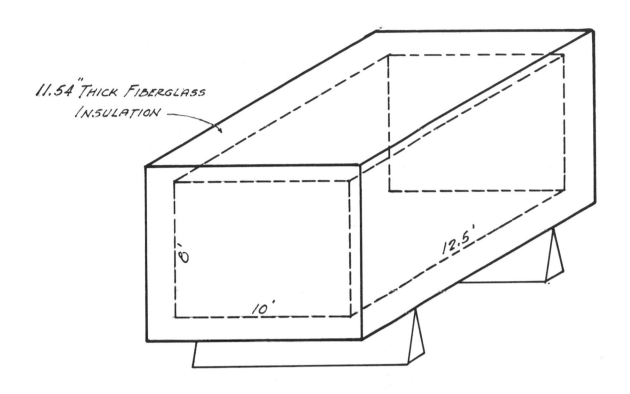

11.54" THICK FIBERGLASS INSULATION

8'

10'

12.5'

INSULATED COMPARTMENT EXPOSED ON SIX SIDES

OCCUPANTS : ESTIMATED 2½ ADULT PERSONS

VOLUME : 8 X 10 X 12.5 = 1000 CU.FT. —400 CU. FT. PER PERSON

VENTILATION : ONE AIR CHANGE OR 1000 CU.FT./HR

HEAT LOSS

ENCLOSURE : 610 SQ. FT. X .0234 X 70 = 1000 BTU/HR

VENTILATION : 1000 X .018 X 70   = 1260
                                     ‾‾‾‾‾‾‾
                                     2260 BTU/HR

HEAT GAIN

2½ PERSONS : RADIATION & CONVECTION   750

                    EVAPORATION   250

4 —100 WATT LAMPS   1260
                                     ‾‾‾‾‾‾‾
                                     2260 BTU/HR

Some very old principles—and some very new ones—were brought to bear on the problem of conservation of heat and cold in living and working spaces. For example, the "Thermos bottle" was, in those days, a homely device familiar to most people. It was used to keep liquids hot or cold; and this was about all it was used for. Then, however, the idea began to attract interest as offering some solutions to the larger problem:

In the interest of conservation as well as economy it behooves us to insulate our buildings as efficiently as possible. This applies in all climates, hot or cold, as it concerns both heat gain and heat loss. When the average annual temperature of a region is 55°F or thereabouts, heating is required, and where it approaches 70°F, cooling is needed. In either case, insulation is the corrective measure, and the cost may be equal to that of the fuel saved.

To date, the evacuated walls of the Thermos bottle are by far the most effective barrier to heat flow, either in or out. And since the greatest heat loss or gain in a building is through the roof or windows, that is where the principle of the thermos could possibly be put to good use. The double glazed T window or Thermopane now available represents a worthy move in that direction, but there is still a further need for improvement, perhaps with crowned glass and a partial vacuum.

In recent years insulating buildings has become standard practice. However, in the light of increasing costs of fuel and power for heating and cooling, much more progress has to be made, or life in hot or cold climates will become hard and economically untenable.

One way to better insulate a roof is shown in the drawing. Lightweight vacuum panels are laid on the joists of the top ceiling, with the butt joints taped. The efficacy of such panels is unknown, but the vacuum plus a bone dry panel provide an impenetrable heat pocket for re-radiation, and stands fair to outclass conventional ceiling insulation consisting of four inch fiberglass bats stapled to the sides of joists. For buildings other than homes, stiffened, 2½-inch thick vacuum panels made of metal-clad plywood can replace the prevailing corrugated steel decking and fiberboard insulation. The ceiling and the roof surface will be flat and smooth because the metal will be drawn in by the suction within the panels.

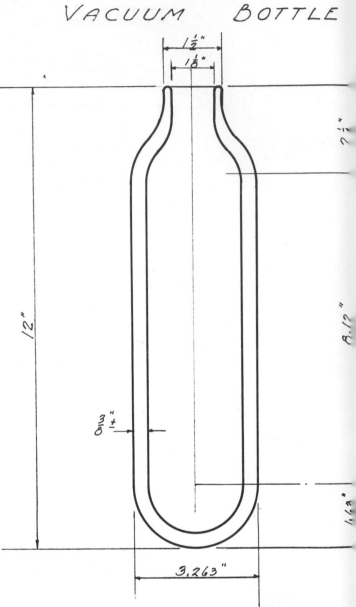

VACUUM BOTTLE

CAPACITY : 2.094 POUNDS OF WATER
TEMPERATURE : 212° INSIDE — 44° OUTSIDE
RANGE 168°F, MEAN TEMPERATURE 128°F
HEAT CONTENT 2.094 x 168° (SENSIBLE) = 351.79 B
TIME TO REACH EQUILIBRIUM — 72 HOURS
HEAT LOSS — 351.79 ÷ 72 = 4.88 BTU./HR.
OUTSIDE SURFACE OF BOTTLE .8108 SQ.FT.
RATE OF HEAT LOSS = 4.88 ÷ .8108 = 6 BTU./HR./FT

ISOTROPIC HABITAT

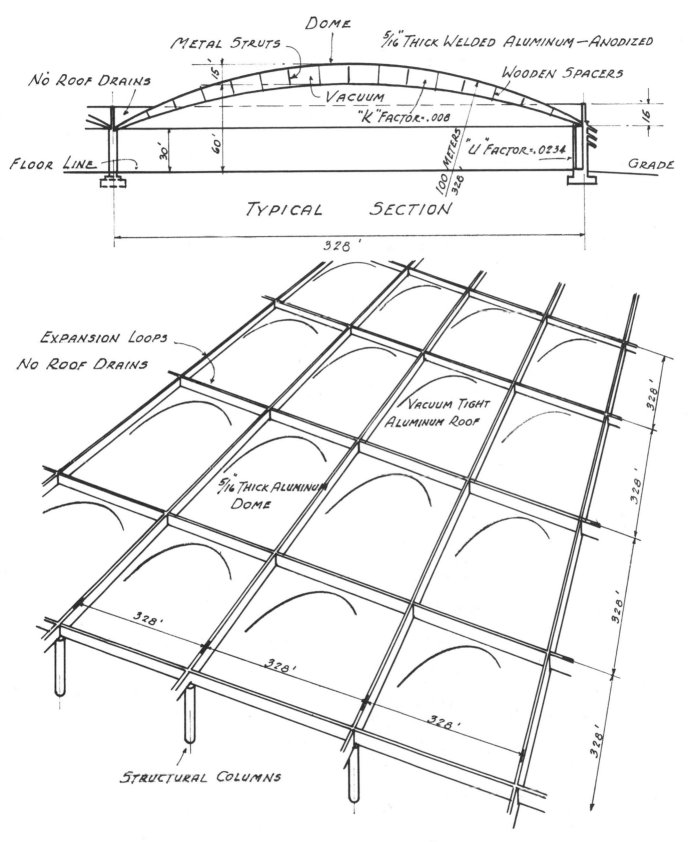

NO ROOF DRAINS

METAL STRUTS

DOME

5/16" THICK WELDED ALUMINUM—ANODIZED

WOODEN SPACERS

VACUUM

"K" FACTOR = .008

"U" FACTOR = .0234

15'

100 METERS / 328'

FLOOR LINE

30'

60'

16'

GRADE

328'

TYPICAL SECTION

EXPANSION LOOPS
NO ROOF DRAINS

VACUUM TIGHT
ALUMINUM ROOF

5/16" THICK ALUMINUM
DOME

328'

328'

328'

328'

328'

328'

328'

STRUCTURAL COLUMNS

ISOMETRIC ROOF PLAN

13

HERMETIC CURTAIN WALL—
CONVOLUTED EXPANSION JOINTS

As a consequence of the impending shortage of fossil fuels, and the dire need for greater economy and efficiency in construction, an updated re-evaluation is in order. This applies particularly to high rise buildings. The common practice has been to build the exterior walls of stone, brick, and concrete or artificial stone. And in recent years many important buildings were built of metal and/or of glass. Unfortunately, none of those are acceptable under the mandatory requirements in this crucial time. For example: The structural elements should be strong, durable and protected against fire. The exterior walls must be safe, light in weight, non-absorptive, leakproof and easy to maintain. The overall coefficient of heat transmission ("U" value) should not exceed .075 Btu per square foot per hour, per degree difference in temperature, inside to outside.

The accompanying drawing offers a new concept in metal curtain wall design. It consists of sheet metal pans of the proper gage and quality, supported by fireproofed structural steel, and backed up by insulation far better than in the past. The edge seams between the pans, including the junctions of vertical and horizontal seams, are welded, so as to make the entire solid facing weathertight against wind and storm and updraft. The metal facing is safe against earthquakes, and will not come apart in case of fire. The pans are stiffened to insure a good appearance, and those on the sunny side of the building are equipped to serve as solar heaters, or evaporative coolers in some localities. The key to this scheme is in the deep longitudinal "V" joints and the light gage flexible intersections, which accommodate both expansion and contraction. This can be a boon to construction in the Pacific coastal region and elsewhere where earthquakes are a threat and, of course, metal buildings are quite certain to hold sway in the future.

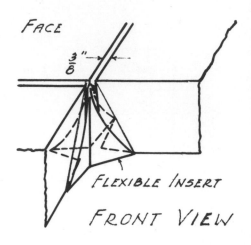

FRONT VIEW

EXPANSIBLE INTERSECTION

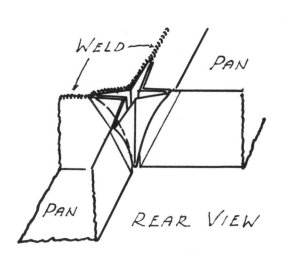

REAR VIEW

JOINT DETAILS

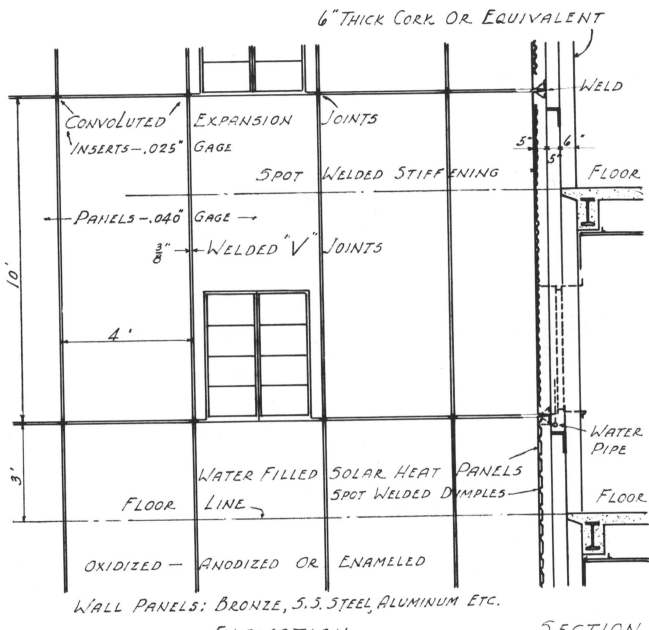

6"THICK CORK OR EQUIVALENT

WELD

CONVOLUTED EXPANSION JOINTS

INSERTS—.025" GAGE

5"   6"

5"

FLOOR

SPOT WELDED STIFFENING

PANELS—.040" GAGE

$\frac{3}{8}$" WELDED "V" JOINTS

10'

4'

WATER FILLED SOLAR HEAT PANELS

WATER PIPE

3'

SPOT WELDED DIMPLES

FLOOR LINE

FLOOR

OXIDIZED — ANODIZED OR ENAMELED

WALL PANELS: BRONZE, S.S. STEEL, ALUMINUM ETC.

ELEVATION   SECTION

15

The Apollo Program—the early venture we all learned about in our Space History courses—came under a great deal of criticism. Its opponents contended that the billions spent on the space program were, by and large, wasted on a useless, though spectacular, series of explorations of the near solar system. But within a few years it became clear that one of the spin-offs of Apollo was the basis for a giant step forward in conservation.

The astronauts who traveled into space required, of course, absolutely airtight containers. The principles used in protecting these space travelers—combined with the ideas that went into the design of the Thermos bottle—ultimately led to a new concept in building:

The accompanying illustration shows a structure with a free-form, waterproofed concrete shell exterior;

it also has vacuum insulation with an effectiveness approaching that of the Thermos bottle and the incandescent lamp.

The inner and outer membranes of the exterior are intended to be made vacuum tight by fusing or sealing the joints or seams with high-frequency electric-induction heating. The rigid fiber glass insulation inbetween the sealed membranes becomes devoid of moisture and air due to the vacuum or negative pressure, and is consequently a much better insulating material, besides minimizing the transmission of sound. The rigid insulation also provides backing to prevent collapse within the evacuated space. The efficiency of the insulation will be proportional to the amount of vacuum that can be held, and this of course depends on the quality of membranes and construction. Vacuum valves should be placed in convenient locations.

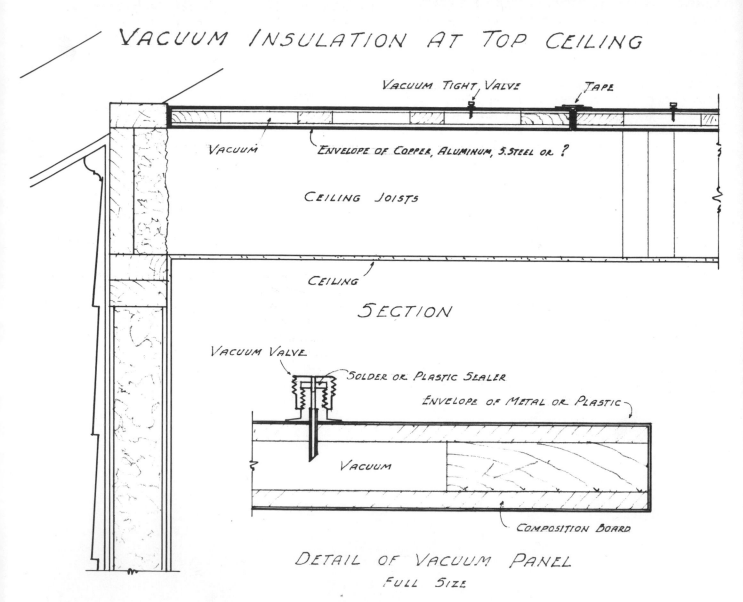

VACUUM INSULATION AT TOP CEILING

VACUUM TIGHT VALVE   TAPE

VACUUM   ENVELOPE OF COPPER, ALUMINUM, S.STEEL OR ?

CEILING JOISTS

CEILING

SECTION

VACUUM VALVE

SOLDER OR PLASTIC SEALER

ENVELOPE OF METAL OR PLASTIC

VACUUM

COMPOSITION BOARD

DETAIL OF VACUUM PANEL
FULL SIZE

WATER IN ABUNDANCE

Today we take for granted the fact that water for drinking and cooking comes out of the *blue* faucet, while water for most other purposes comes out of the *orange* one. Most of us never question this arrangement, and many younger people would be unable to explain why we have two kinds of water.

But the concept of two-source water supply—now commonplace in metropolitan areas around the world—is quite a recent development. It has been established as a result of the concern that grew in the decades 1960–1980 about the scarcity of potable water. During that time people were growing very concerned about their water supply. Typically, *all* of the water in household use came from one source—the single-reservoir system. This water was subjected to increasingly massive doses of chemicals and increasingly rigorous treatment processes to make it fit to drink. Then, when it was piped into the consumption area, the highly-purified water was used for bathing, washing, laundry, sprinkling lawns, hosing down streets, heating, cooling, fighting fires; in short, for everything. Occasionally, during dry periods, the use of water was severely restricted.

Government agencies made studies and encouraged inventors to find ways to make drinking water from seawater. Pilot desalinization plants were established; but the cost remained prohibitive.

People looked again at the immense amount of fresh water that was wasted under the existing system. The following are notes and drawings of the time, exemplifying the beginning of a shift in thinking that has led to the water system we enjoy today:

Serious shortages of water will increase in frequency and will become critical in the years ahead unless we can devise a better way to meet our needs. This is a proposal for the assurance of an adequate water supply to meet the exigencies of municipal growth. There is ample fresh water. We waste most of it.

A more or less typical example is illustrated by the table below. The setting is an industrial city of 100,000 people in New England. The terrain is hilly, and a small river passes through the town. The watershed beyond the city limits is rural and sparsely inhabited.

NATURAL WATER

Watershed	246 square miles
Area of city	28.2 "
Annual rainfall (evenly distributed)	45.53 inches
Overall source of water from precipitation	534,066,432 gallons per day
Measured outflow at river	309,571,200 " " "
Lost to evaporation to the substrata and to the growth of all living things (42%)	224,495,232 gallons per day
Rainfall within city	60,883,573 gallons per day
Outflow	35,291,116 " " "
Lost to evaporation, etc.	25,592,457 gallons per day

CITY WATER
(From Reservoir)

Average daily use	14,000,000 gallons
(6,500,000 high pressure 200 to 300#	7,500,000 low pressure) 130#

Proportion

Food and drink	500,000 gallons	3.5%
Other uses in the home	4,500,000 gallons	32.
Industrial and general	9,000,000 gallons	64.5
	14,000,000 gallons	100 %

Heavy industry makes some use of lake and river water.

These figures show that a very small percentage of our water supply is used for food and drink, and that much of this water which we pamper and coddle and guard so carefully is used for purposes where water of a lesser quality would be satisfactory. Common water from lakes and rivers—screened, filtered, and sterilized—would be suitable for washing, bathing, laundering, sprinkling lawns and streets; and for heating, cooling, fighting fires, and for process water in industry. Hence a second source of water recommends itself.

We treat all of our water to make it fit to drink—and yet we use less than 5% of it for this purpose. This does not make sense. Instead, we should think about a two-source system.

Source one would be potable water. This water would deserve the most careful attention. It would come from the highest and purest reservoirs and receive the utmost in treatment. At present we fluoridate and chlorinate our water. Let us concentrate the full force of advanced scientific knowledge on this precious element of our water supply. We could do everything possible to make our drinking water softer, sweeter, clearer, and more healthful. Perhaps the benefits of the rare mineral waters could be made available.

Obviously this improved water would command a higher price; but we would be using much less of it, since it would be used only for drinking, cooking, and, perhaps, bathing.

Source-two water would be made available from lakes, rivers and flood control basins. This water would be properly treated for safety—but it would not receive the advanced treatment accorded to Source-one water. Water from source two would then be made available—at much lower price—for irrigation, sewage disposal, lawn-care, street-washing, industrial applications, and a host of other uses.

This dual water supply would involve a double system of piping. The initial cost would admittedly be high—but not as high as might appear at first glance. The source-one piping would be smaller and cheaper; and the source-two piping and fittings, although larger, could be of lesser gauge and strength because of the lower pressure and the inert nature of the water.

What about the dangers of cross-connection—introducing source-two water mistakenly into situations where it would be drunk? Revised plumbing codes would specify that the piping be absolutely separate and that the potable water be unmistakably identified by bright and luminous color or other effective means.

In the long run the two-source water system will bring us the best water in amounts sufficient to meet our eating and drinking needs. It will also go far toward assuring following generations a good and ample water.

As we know, the two-source water supply is a staple of our daily living in the 21st century. The next time you turn the blue tap and draw a glassful of clear, sparkling water, it may come to mind that this was not always something that could be taken for granted.

TWO SOURCE WATER PIPING

SCHEMATIC ELEVATION

UTILIZING RIVER WATER

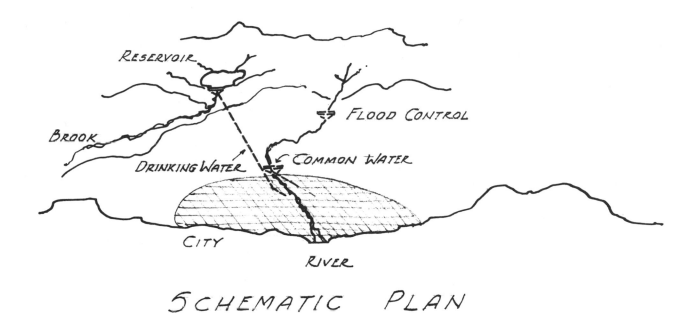

RESERVOIR

FLOOD CONTROL

BROOK

DRINKING WATER

COMMON WATER

CITY

RIVER

SCHEMATIC PLAN

RAINWATER FOR IRRIGATION AND POWER

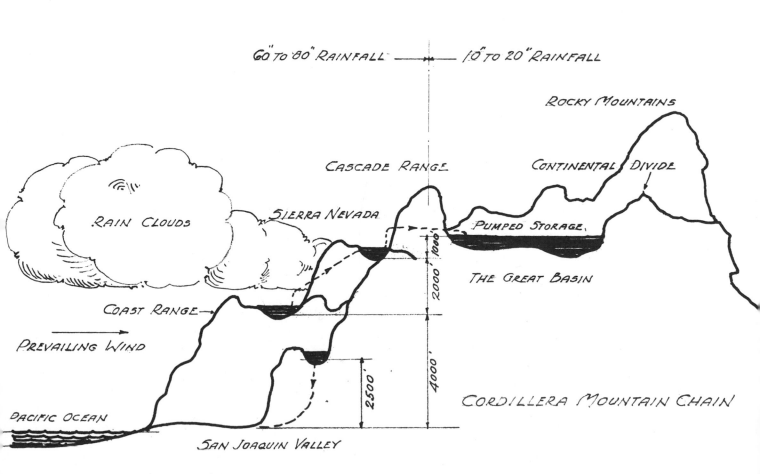

60" TO 80" RAINFALL ——— 10" TO 20" RAINFALL

ROCKY MOUNTAINS

CASCADE RANGE

CONTINENTAL DIVIDE

RAIN CLOUDS

SIERRA NEVADA

PUMPED STORAGE

1000'

2000'

THE GREAT BASIN

COAST RANGE

PREVAILING WIND

4000'

2500'

CORDILLERA MOUNTAIN CHAIN

PACIFIC OCEAN

SAN JOAQUIN VALLEY

SECTIONAL ELEVATION

RECLAMATION OF KITCHEN
AND HOUSEHOLD WASTE

The book you hold in your hand is a product of reclaimed waste. The containers in which we buy our food, the fences and walls surrounding our grounds, the implements we use around the house—all may be, at least in part, the results of waste reclamation.

There was a time when all waste was simply thrown away. As recently as the last decade of the previous century people were still resorting to what was called "landfill"—combining all waste and garbage, dumping it in a hole, and covering it with earth. Sand, bottles, and papers were strewn along the countryside. Everyone talked about the problem of what to do with the garbage, but little was done.

One problem was human attitude. People would not go even to the trouble of separating household waste before disposal so that it could be compacted and perhaps reclaimed.

Now, of course, practically every kitchen has its WasteSort unit, either manual or power-driven. We are use to dropping our refuse into the appropriate bin of the three constituting the WasteSort's console; and we know that the waste will go to the municipal recovery area for reclamation, disposal, or treatment.

However, it was not so long ago that such a system was considered impossible to attain. Only when people were confronted with mountains of garbage—with no place to put it—was there a serious effort to put into practice the theories of those who had been warning of the situation for some time.

Here is one of these warnings, drawn from the mid 1970s—along with suggestions and illustrations that you may recognize as being the somewhat primitive forerunners of our present systems:

Truly a change in waste disposal seems imperative. Ultimately all waste material, both organic and mineral, will probably be utilized and reprocessed to yield multiple benefits. Such an undertaking requires organization, and properly managed chemical, metallurgical, and manufacturing plants. This may be slow in coming—so until then, let us begin by washing, separating, and compacting the materials such as tin-cans, bottles, and papers; also scrap iron, aluminum, copper and brass, and plastics. Each group can be deposited on unfertile ground or in bins, where it can be retrieved whenever it becomes profitable. This can be done as a community operation or singularly by the individual family.

ECOVERY FROM HOUSEHOLD WASTES AND SCRAP

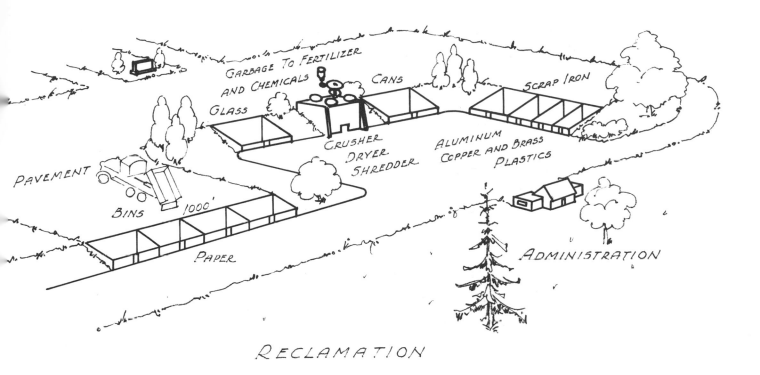

GARBAGE TO FERTILIZER
AND CHEMICALS
GLASS
CANS
SCRAP IRON
CRUSHER
DRYER
SHREDDER
ALUMINUM
COPPER AND BRASS
PLASTICS
PAVEMENT
BINS 1000'
PAPER
ADMINISTRATION

RECLAMATION

GARBAGE TO FERTILIZER

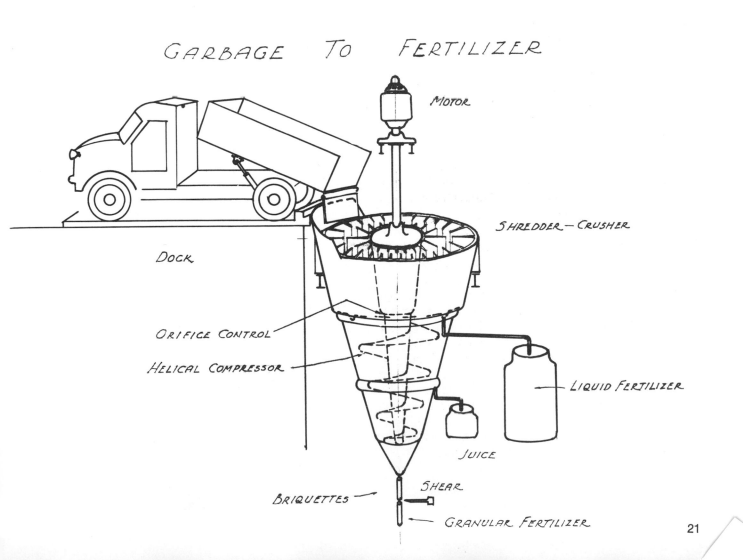

MOTOR

SHREDDER — CRUSHER

DOCK

ORIFICE CONTROL

HELICAL COMPRESSOR

LIQUID FERTILIZER

JUICE

BRIQUETTES

SHEAR

GRANULAR FERTILIZER

21

NOURISHMENT FOR THE EARTH

Even in our cities we enjoy the sight of green grass and growing things. Some of us may know—most of us do not—that our soil has been made far more productive through the intelligent use of certain simple, natural materials that were, at one time, thrown away. These materials are available in our familiar composting units; the small individual ones that dot the suburban landscape and the larger structures that may be seen at the regional composting centers.

You may be interested in the brief excerpt that follows, drawn from the files of a newspaper in a small mid-Atlantic state community, published in 1972:

There are several kinds of unnecessary waste, each attributable to lack of wisdom or knowledge. Prominent among them is the burying of airtight plastic trash bags filled with leaves, grass clippings, weeds, sod, brush, and prunings. Such practice is simply contrary to nature, and in effect is postponing an evil day.

To dispose of these materials should be a community responsibility, and should be done in a scientific manner. A compost farm could be set up to yield nutritious compost, humus, and loam, for sale in a self-supporting venture. A similar enterprise in which sod is produced is already in existence. To equip such an operation requires special trucks for mechanical loading, shredding, and compacting; also tools for spreading sandy soil, and agricultural lime. Manure and/or bacterial activators are added separately as desired.

This concept is a reaffirmation of the principles expounded in Faulkner's "Plowman's Folly." It was put into practice, in a certain direction, by Rhodale in Emmaus, Pennsylvania, and by others who are growing health foods by organized gardening, using a limited amount of commercial fertilizers, if any.

COMPOSTER

STAINLESS STEEL OR BRONZE

GRASS CUTTINGS
LIME
DIRT

LEAVES

WELDED WIRE MESH
6 FT. DIA. ROUND

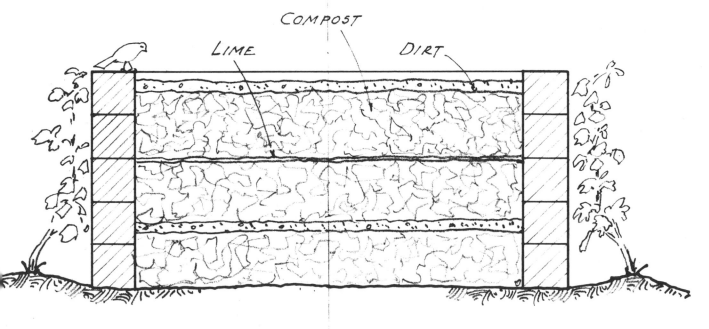

COMPOST
LIME
DIRT

MANHOLE BLOCKS
ROUND OR ELLIPTICAL

SEWAGE RECLAMATION

Sometimes as we travel at night we see, far away, the flickering flames of the Disposal Sites. We know in general that these areas have something to do with the handling of sewage; and then we forget about the topic.

We are able to dismiss the thought so easily because we have no worries about the efficient and unobtrusive dissolution of human waste. However, a generation or two ago this was an annoying and, in many ways, a dangerous concern.

In the last quarter of the 20th century the treating and disposing of human waste had shown minimal progress. In rural and suburban areas the septic tank—basically a local receptacle with a leaching field or dry well—was in wide use; and there were increasing instances of backup and flooding. In urban areas sewage was pumped to treatment plants where it was reduced to a relatively harmless state. There were some more ingenious schemes; notably in Milwaukee, Wisconsin, where human sewage was turned into fertilizer and sold throughout the United States.

The essential idea of the movement of human waste was flushing with water, so that the waste material flowed easily through pipes running about 1/3 full. This flushing involved great volumes of potable water, because the two-source water supply had not yet been instituted.

Certain observers of that time felt that something better could be done. Here is a brief contemporary description of such an approach:

We need a better solution to waste disposal problems. These drawings and notes will, it is hoped, lead toward such a solution.

One drawing shows a saucer-shaped lagoon for the fermenting of raw sewage on level land. At the center, the earth should be removed down to bed rock, allowing water to flow through the accumulated sewage, down into the fissures. An airtight cover of oil or a plastic film, plus the warmth of sunlight augment the putrefaction and degeneration of the materials. The gases that develop are collected in floating cones, and are automatically ignited and burned. A surrounding wall excludes outside surface water and sends any possible odors skyward.

The second drawing is for hilly or mountainous country. The digester can operate similarly to the first on, but faster decomposition and neutralization may be possible due to its greater depth and pressure. Moreover, offensive odors are less likely because of the smaller surface area exposed to the air, and the stratification.

The disposal system we now enjoy (and much of our enjoyment lies in the fact that we have to think about it so little) is more elaborate than the idea put forth in these notes; but the principals are the same.

SEWAGE

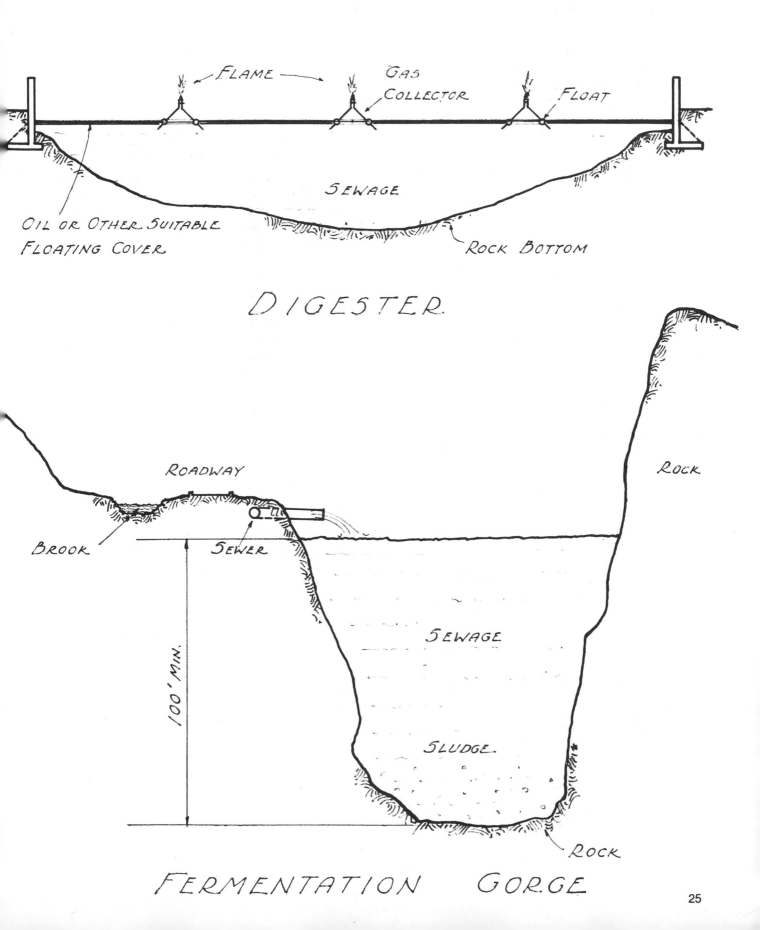

FLAME

GAS COLLECTOR

FLOAT

SEWAGE

OIL OR OTHER SUITABLE FLOATING COVER

ROCK BOTTOM

DIGESTER.

ROADWAY

ROCK

BROOK

SEWER

100' MIN.

SEWAGE

SLUDGE.

ROCK

FERMENTATION GORGE

The Power of Moving Air

The wind is one of the oldest sources of power—and yet one of the newest. Throughout most of the 20th century the windmill was considered a quaint anachronism. Only when people began to see the contrast between the pollution from fossil fuels and the cleanliness of power from the air currents, did they begin to think about returning to this "primitive" source of energy. You may be interested in the following devices used in the late 20th century to again harness the power of the air. They are the forerunners of our present wind-energy systems.

PROPELLER-TYPE WINDMILL

The propeller-type windmill is a standard. It has functioned reasonably well, but its coefficient of performance leaves room for improvement. The flat design has the advantage of being exposed to the full force of the wind in its entirety; though less than 50 percent of the face of the mill is hit by the wind. And of course the angle of incidence will also reduce the effectiveness of the remainder by another 43%.

For the purpose of design, a wind velocity of sixteen to twenty miles per hour is considered reasonable. The diameter of wheel may range from ten to about twenty five feet for best results. Under design conditions the wheel would revolve sixty-five times per minute for the ten-foot size, down to thirty times per minute for the twenty-five-foot size. A windmill will receive a pressure of about one pound per square foot from a sixteen mile an hour wind, and four pounds per square foot from a thirty-two mile per hour wind. The principle involved is that of Baker's mill, which demonstrated that, "every action is always opposed by an equal reaction." Some time later this became Newton's third law in physics.

The propeller-type windmill can be improved by providing self-lubricating bearings, and a frame of rustfree metal. Those constructed of wood, as in Europe, were difficult and costly to maintain, while those that were popular in this country suffered because of neglect in regard to lubrication and protection of the rustable metal frame.

The windmill was useful in the earlier days for pumping water into storage tanks for use in the household and in the barns. This can continue, and in addition the windmill can generate electric power and charge storage batteries for electric automobiles and other uses.

There are drawbacks to wind power, in that it is not constant, and it occurs all odd times—not just during working hours. Until we learn how to store energy more efficiently we will have to be content with using windmills for pumped water storage, hydraulic accumulator storage, high pressure air bottles, and electric storage batteries. To forsake or ignore wind power would be foolish, but to use it whenever available, to store the energy as best we can, and to supplement this energy or power with fuel generated electricity when necessary, is practicable and economical.

SOLAR — AMPLIFIED WINDMILL

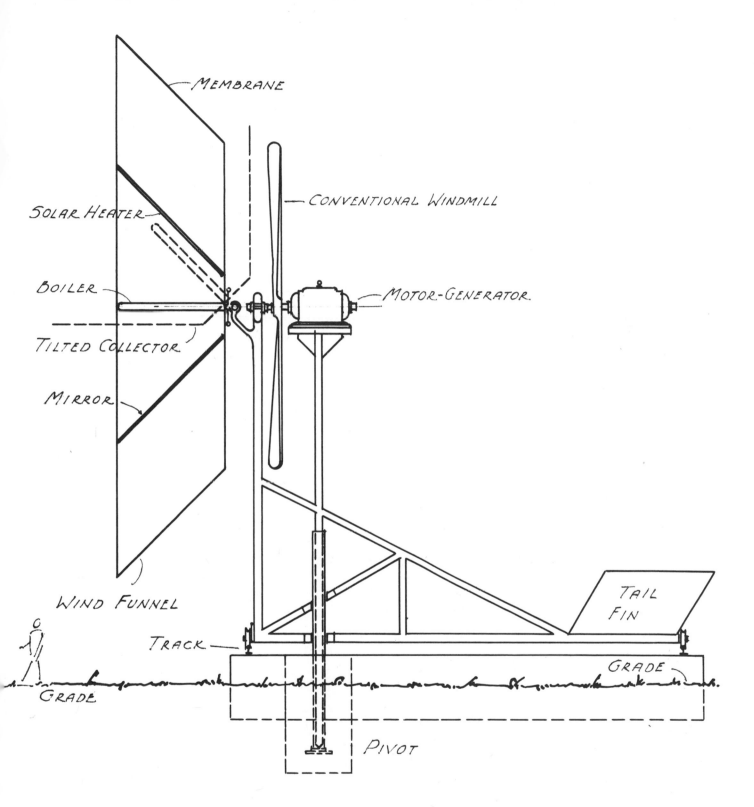

MEMBRANE

CONVENTIONAL WINDMILL

SOLAR HEATER

BOILER

MOTOR-GENERATOR.

TILTED COLLECTOR

MIRROR

WIND FUNNEL

TAIL FIN

TRACK

GRADE

GRADE

PIVOT

ELEVATION

REVOLVING WINDJAMMER

Windmills should be placed on promontories, on high ground, and in geographical locations where there is an abundance of either forced or induced air flow. In designing a windmill there are several important things to consider. The more vital elements of good design are: durability, lightness of the moving parts, frictionless bearings; as well as determining the working range of wind velocity, and utilizing or controlling violent wind action as in heavy gusts or tornadoes.

The drawing shows a balanced double-action or revolving windmill. It consists of two partly-shielded paddle wheels with vertical axes, a wind splitting or diverting structure, and the geared transmission of power to a generator; with the entire assembly riding on a circular track and pivots. The exact design and specification depends on data concerning local wind and weather conditions, and the size of the installation.

REVOLVING WINDJAMMER

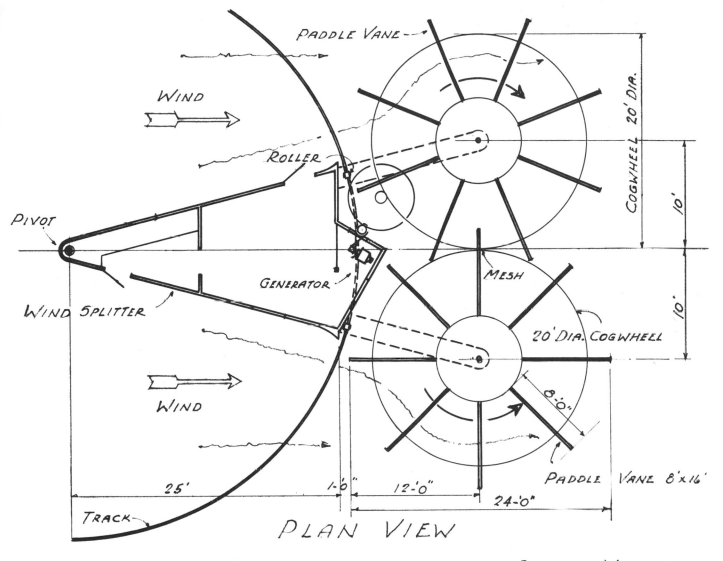

PADDLE VANE

WIND

COGWHEEL 20' DIA.

10'

ROLLER

PIVOT

GENERATOR

MESH

WIND SPLITTER

20' DIA. COGWHEEL

10'

WIND

8'-0"

25'

1'-0"   12'-0"

PADDLE VANE 8'×16'

24'-0"

TRACK

PLAN VIEW

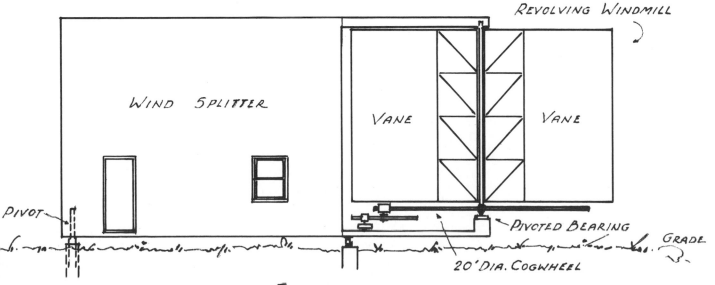

REVOLVING WINDMILL

WIND SPLITTER

VANE

VANE

PIVOT

PIVOTED BEARING

GRADE

20' DIA. COGWHEEL

ELEVATION

WIND POWER CONCH

Winds range from a zephyr to a tornado; from causing a ripple on water to creating a wave with a destructive force equal to that of a violent explosion. The extremes are of little interest so far as available power is concerned, but in between, and in certain localities, winds of considerable velocity prevail much of the time, from which a goodly amount of energy may be had. (Obviously a substantial amount of investment is involved).

The conventional anemometer, having horizontal whirling cups, can utilize a large portion of the total force of the wind. One at Mount Washington finally fell apart as it clocked winds of 243 miles an hour.

The most essential considerations in the design of any wind power equipment are: lightweight construction; a minimum of friction; and, of course, a favorable location. One important objective is to find the optimum of tolerable resistance in extracting power at the various design velocities. If the resistance of the moving parts is excessive, too much of the wind will by-pass or circumvent the mechanical power generating apparatus, resulting in a reduced level of efficiency.

As a note, the amount of available power from wind and from deep sea waves is practically alike, though the force of water in motion is much easier to trap than that of wind. This is due to the greater mass and inertia that holds the force of water in a forward direction. Conversely the force of the air with a mass of only 1/800 that of water, will easily divert or change its course, unless it is driven at high velocity. As a practical illustration, a five-foot ocean wave caused by twenty mile winds will exert an upward force of 1932 foot-pounds per square foot per minute—while a direct wind of twenty miles per hour can yield 2283 foot-pounds per square foot per minute.

WIND POWER CONCH

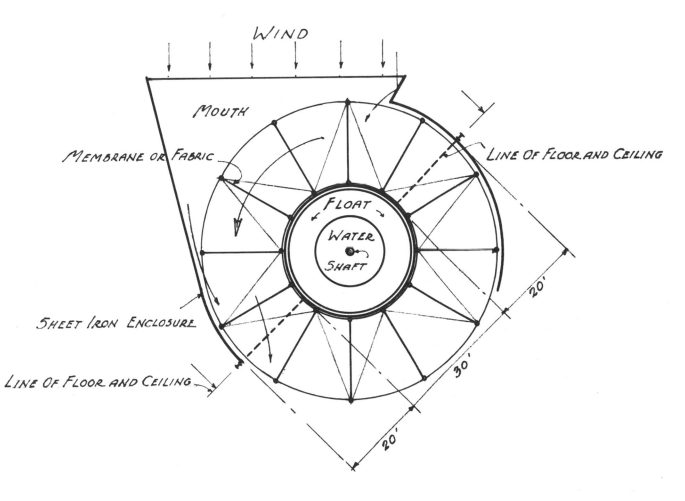

WIND

MOUTH

MEMBRANE OR FABRIC

LINE OF FLOOR AND CEILING

FLOAT

WATER

SHAFT

SHEET IRON ENCLOSURE

LINE OF FLOOR AND CEILING

20'

30'

20'

PLAN

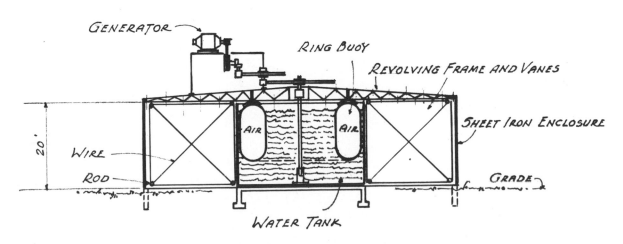

GENERATOR

RING BUOY

REVOLVING FRAME AND VANES

AIR

AIR

SHEET IRON ENCLOSURE

20'

WIRE

ROD

GRADE

WATER TANK

SECTIONAL ELEVATION

SCHEMATIC

BARREL-TYPE WINDMILL

Wind is the first, or at least the most evident moving force caused by the heat of the sun. The idea of harnessing the power of the wind has been given some thought, but as yet it has brought about few inventions that might be useful in the modern world. Heretofore thinking was mainly in terms of horizontal wind pressure, as for sail boats and propeller-type windmills, and hardly any thought was given to the power of vertical airflow, or the updrafts in mountainous areas. Nor has there been much progress in sapping the destructive energy of tornadoes and hurricanes.

In our era, covering about one hundred years of abundant supplies of coal, gas, and oil, the cost of power has been so cheap as to cause the familiar Dutch windmills with cloth-covered arms, and the steel-bladed windmills of American farms to disappear from the scene. This disappearance represents a dip in the experience curve of wind power machines; but there is hardly any doubt that the curve will sweep upward as the price of fuel inevitably rises, and as invention and ingenuity come into their own.

The drawing illustrates a deviation from the propeller-type windmill to the barrel or squirrel-cage type. Absorbing the head-on force of the prevailing wind in this manner should develop an improved coefficient of performance, and also permit lightweight construction, which includes vanes or panels covered with fabric or synthetic sheet materials. Locations such as Bigelow Lawn on Mount Washington, along the Oregon coast, and along the Netherlands coast are especially well suited for this type of wind power installations.

BARREL TYPE WINDMILL

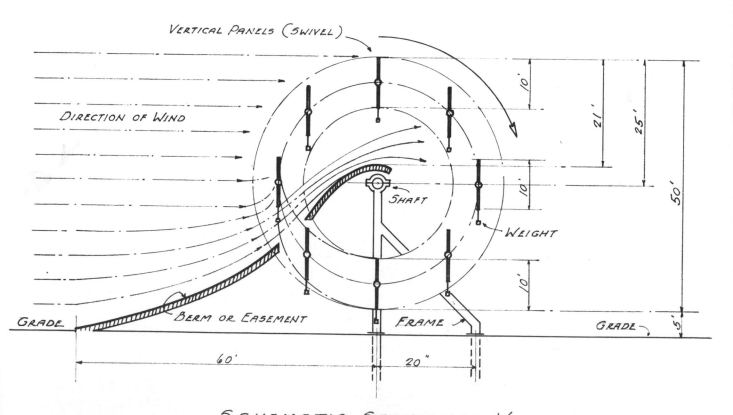

SCHEMATIC SECTIONAL VIEW

A WIND-POWERED FOREST RANGER STATION

Forest fires present a constant hazard. Fire-watch stations are remote, and difficult to reach with electric and telephone lines.

Here is an application of wind energy that powers the forest ranger station and also serves as a lightning rod. The vanes, turned by the wind, operate a generator in the base of the installation, providing light, heat, and power for a radio. The shaft is grounded to act as a lightning arrester. An auxiliary unit is available for stand-by power.

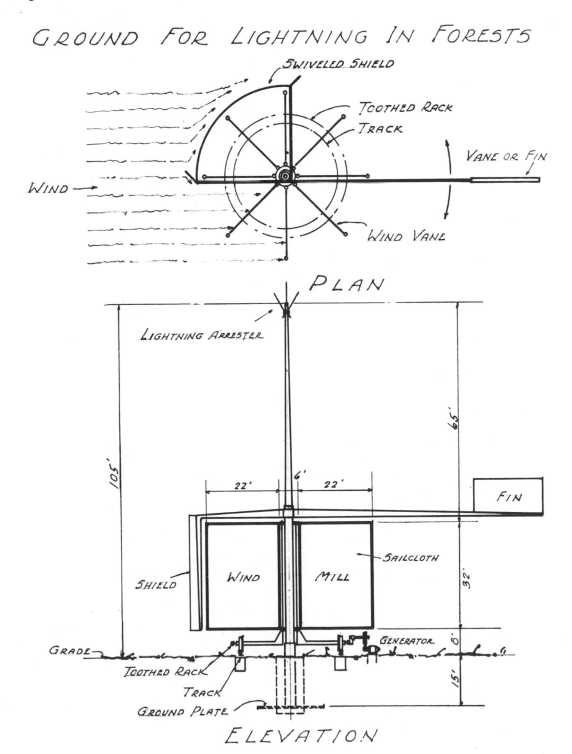

GROUND FOR LIGHTNING IN FORESTS

SWIVELED SHIELD

TOOTHED RACK

TRACK

VANE OR FIN

WIND →

WIND VANE

PLAN

LIGHTNING ARRESTER

105'

65'

22' 6' 22'

FIN

SHIELD   WIND   MILL

SAILCLOTH

32'

GRADE

GENERATOR 6'

TOOTHED RACK

TRACK

15'

GROUND PLATE

ELEVATION

Trapping the Sun's Energy

The sun is our most ever-present and obvious source of energy. But until the last few decades the sun was inaccessible as a means of generating specialized and concentrated power. During the years 1960–1980 "solar houses"—structures with slanted glass roofs, heat-absorption panels, and heating coils—were built in certain tropical and semi-tropical areas, but these buildings were largely regarded as curiosities rather than as prototypes.

Then came a number of developments that combined to bring Apollo's Chariot closer to the reach of earth-bound men. Some of these developments grew out of the early ventures in space travel that took place around that time. For example, the American NASA project at Waltham, Massachusetts, announced a new, fast way of growing the silicon crystals needed for conversion of the sun's energy. For the first time the use of such crystals became practical.

As power from the sun began to emerge from the pages of science fiction, attention was focused on the experiments of such pioneers in solar energy as Dr. C. G. Abbott of the Smithsonian Institution. As early as 1940, Dr. Abbott had developed a *solar boiler,* consisting of a parabolic trough with a water tube boiler and a steam chest.

French scientists also began to experiment with methods of collecting and using the sun's rays. Their first structures, built in the Sahara Desert, culminated in 1960 with the building of a solar furnace on Mount Louis in the Pyrenees. This device, a dished parabolic collector, was sixteen meters in diameter. Later French efforts led to a more sophisticated solar furnace, built near Ordeillo. This installation is still in use for uncontaminated metallurgical testing. A triple-throw system of mirrored reflectors amplifies solar heat to 6,300°F.

At this point let us review some of the thinking that was prevalent in the sunrise of the era of solar power. Some of the following ideas we recognize as only having emerged into full fruition in the 21st century. Others are significant only because they led on to further breakthroughs in the exploitation of the power of the sun which has meant so much to the world.

First, the rays of the sun must be collected. There are four practical methods: the parabolic saucer, the cone, the parabolic trough, and a modified double- or triple-throw long-range reflector somewhat similar to the one in Ordeillo, France.

For ease in fabrication, transportation, and erection the size of units should lie within a fifty-foot dimension. The reflectors can be made of sheet steel lined with mirrors, sheet aluminum, sheet copper, stainless steel, or enameled iron.

All reflectors must be precision-made. The parabolic saucer is the most difficult to construct because it involves stretching the metal in the forming operation. However, this becomes simple with the aid of a suitable press.

The parabolic trough or dished strip is ideal for production on a modern shape-rolling machine, and can be expected to retain its form if the gage and temper of the metal are correct. The trough should not be required to support anything other than its own weight and possibly some wind or wet snow. Nor should it be fastened in any way that will not permit free movement for expansion and contraction.

The cone has no parabolic curves but only radial curves and straight lines, and is therefore the easiest to build. It can be built-up of plates cut to a template, then warped and welded together. The assembly should be easy and accurate with the help of a jig or steel frame, and by clamping, tack welding, and finish seaming.

The long-range reflector can be made up of plates with a slight radial camber in both dimensions. The reflector may be pivoted or on a track, and the plates can be fixed or adjustable depending upon other engineering considerations.

Even now we have several new types of solar heat collectors that we could use.

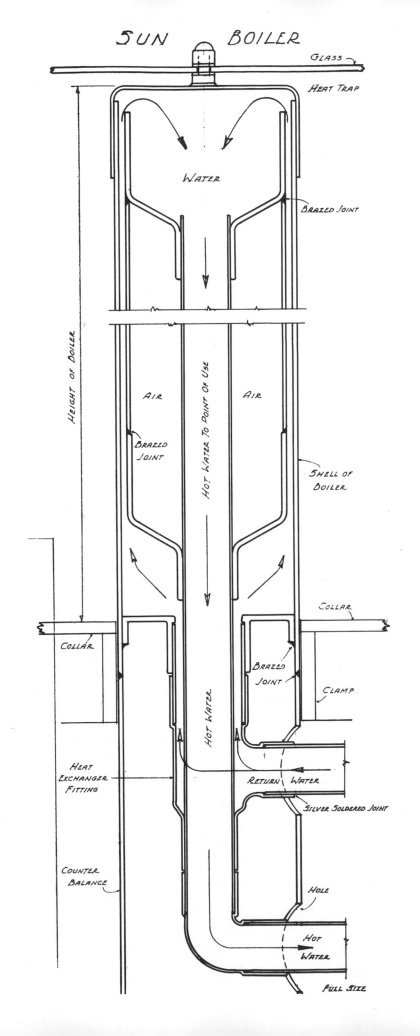

SUN BOILER

GLASS

HEAT TRAP

WATER

BRAZED JOINT

HEIGHT OF BOILER

AIR

AIR

BRAZED JOINT

HOT WATER TO POINT OF USE

SHELL OF BOILER

COLLAR

COLLAR

HOT WATER

BRAZED JOINT

CLAMP

HEAT EXCHANGER FITTING

RETURN WATER

SILVER SOLDERED JOINT

COUNTER BALANCE

HOLE

HOT WATER

FULL SIZE

ALTERNATE SOLAR AND WIND POWER

There is vast potential energy in the sun and the wind, but large scale plant and equipment is required to develop a respectable amount of it. There is also the problem of discontinuity—periods during which there is no sunshine or wind. Storage of energy presents another practical difficulty.

However, it is possible to double the favorable probabilities by building an apparatus that can draw power from either the sun or the wind. Under optimum conditions, such equipment can serve throughout a 24-hour day.

One application of this idea involves a trough-type solar heater with mirrored reflectors. The reflectors multiply the effects of the direct solar rays. They also form a funnel for guiding the wind into the elongated cups of a turbine type windmill. The entire assembly rotates horizontally to track the sun, and also to point the funnel into the wind.

Another device for using both sun and wind is the solar cone wind whorl. This is a helix, made of sheet metal, which will hold its shape in wind and storm. It serves both as a conical reflector and a whirling vane.

SOLAR CONE OR WIND WHORL

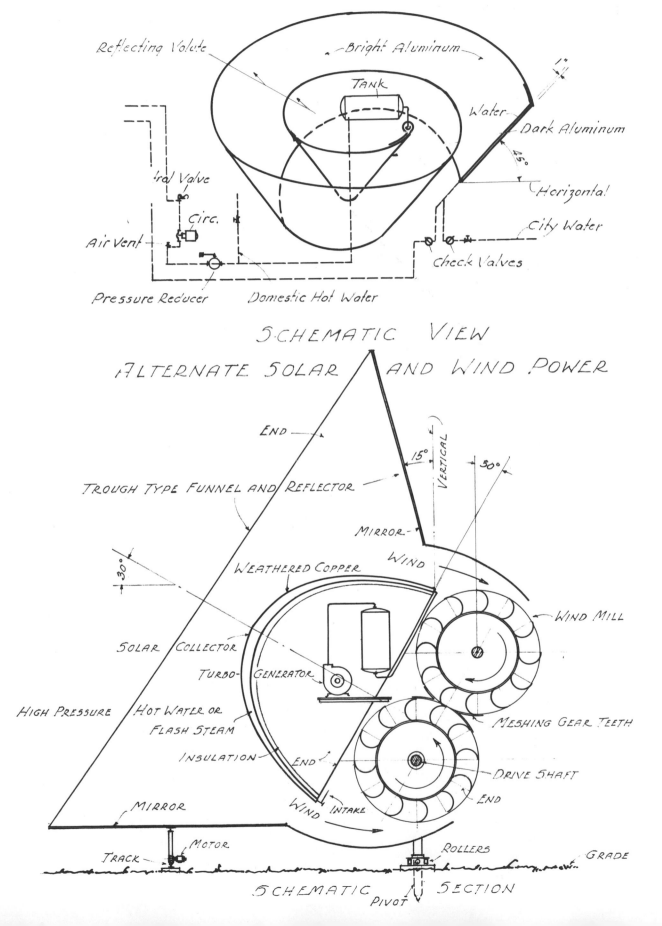

Reflecting Volute

Bright Aluminum

Tank

Water

Dark Aluminum

45°

Horizontal

City Water

ral Valve

Circ.

Air Vent

Check Valves

Pressure Reducer

Domestic Hot Water

SCHEMATIC VIEW
ALTERNATE SOLAR AND WIND POWER

END

15°

VERTICAL

50°

TROUGH TYPE FUNNEL AND REFLECTOR

MIRROR

WIND

WEATHERED COPPER

30°

WIND MILL

SOLAR COLLECTOR

TURBO-GENERATOR

MESHING GEAR TEETH

HIGH PRESSURE

HOT WATER OR
FLASH STEAM

INSULATION

END

DRIVE SHAFT

END

MIRROR

WIND

INTAKE

MOTOR

ROLLERS

GRADE

TRACK

SCHEMATIC SECTION

PIVOT

CONICAL COLLECTOR

The conical collector is a newcomer. It originated with a 60°-cone having the vertex of a similar cone placed inside, in an inverted position to form a boiler. The design advanced to a double cone that could rotate from east to west, on a telescopic mount with its axis parallel to the axis of the earth. The cone has a heat exchanging boiler simulating a pistil standing at the center. A later design envisions a triple cone, with a shorter boiler pistil and higher ratio of collector area to boiler surface area, resulting in quicker action and possibly a higher temperature.

Probably the most sensible type of conical collector is one with a 90° scope and a full-height boiler, consisting of pipes within pipes for efficient heat exchange. The cone can be of heavy gage sheet aluminum polished inside, and the boiler can be of copper tubing with a deep brown oxidized finish. The mouth of the cone is covered with greenhouse-quality plastic sheet (fiber glass) or polyethylene, with the lap seams sealed by means of ultra high-frequency induction heating.

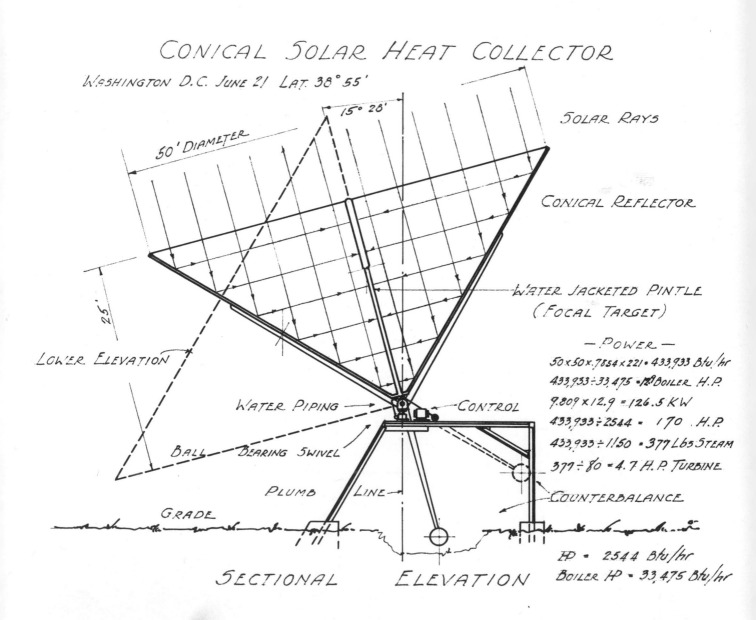

CONICAL SOLAR HEAT COLLECTOR

Washington D.C. June 21 Lat. 38° 55'

15° 28'

50' DIAMETER

SOLAR RAYS

CONICAL REFLECTOR

WATER JACKETED PINTLE (FOCAL TARGET)

25'

LOWER ELEVATION

— POWER —

$50 \times 50 \times .7854 \times 221 = 433,933$ Btu/hr

$433,933 \div 33,475 = 12.9$ Boiler H.P.

$9.809 \times 12.9 = 126.5$ KW

$433,933 \div 2544 = 170$ H.P.

$433,933 \div 1150 = 377$ Lbs Steam

$377 \div 80 = 4.7$ H.P. Turbine

WATER PIPING

CONTROL

BALL BEARING SWIVEL

PLUMB LINE

COUNTERBALANCE

GRADE

SECTIONAL   ELEVATION

HP = 2544 Btu/hr

BOILER HP = 33,475 Btu/hr

LONG-RANGE REFLECTOR/COLLECTOR

The French successes in the Pyrenees have shown that the sun's rays can be reflected hundreds of feet, and with a proper mechanism and fine adjustment strike a relatively small target with accuracy. This uncovers the possibility of generating steam and electricity through the double-throw effect of reflected radiant rays of the sun. This method is suited for power installations of considerable size. The source of energy is steady and timeless, but the amount per unit of reflector area per minute or hour is quite small, and for that reason many relectors and a great deal of construction are involved.

The reflectors may be large or small, whichever prudence, mechanics, and economy dictates. They need to turn to follow the sun from east to west, and to lean backward and forward as the sun rises and lowers in the sky. The reflectors may float, pivot, or run on a radial track, all electronically and mechanically controlled and operated. The collector can be stationary, and preferably concaved, while one or all of the reflectors can aim at the same collector. A solar heat trap can be used in conjunction with the boiler feed water system, raising its temperature to near the boiling point. Further experience and research are needed to make this a competitive system for producing electric power.

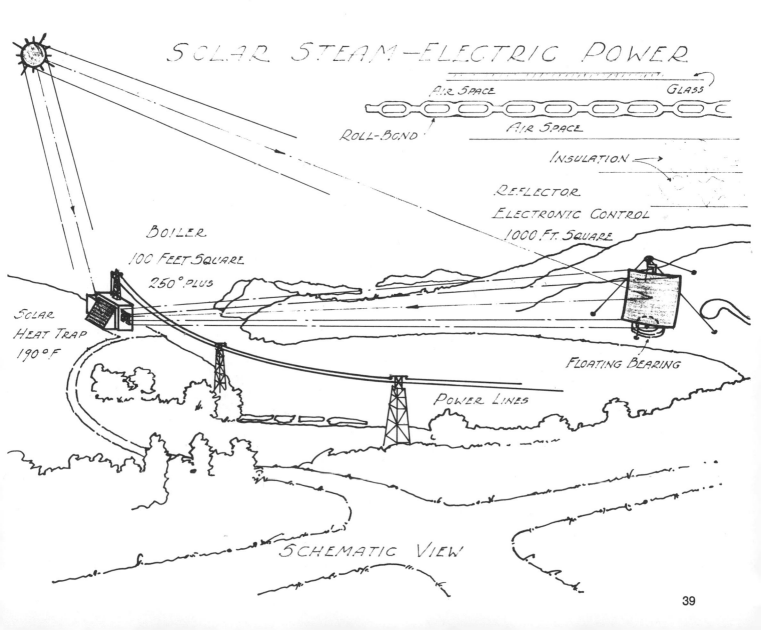

TROUGH-TYPE COLLECTOR

The trough-type collector, as demonstrated by Dr. Abbott, is capable of producing low pressure steam and driving a steam engine. Its reflector/collector ratio can hardly be greater than 12:1, therefore it is necessary to guard against heat loss, using insulation and other means, while delivering the collected heat to the generator.

For warm climates where the sun is high, the air is clean, and the sunshine hours are many the trough-type collector may be more economical than the cone. This system deals in lower operating temperatures, which are usually easier to hold, but in this case the hot pipes over the reflecting troughs are subject to heat loss by convection due to wind and movement of the surrounding air, and some heat loss by radiation. Dr. Abbott experimented with both glass sleeves over these pipes and a partial vacuum.

While the collector pipes should be of copper with silver brazed joints, the circulating pipes leading to and from the power house can be of steel with screw or welded joints, and must be well insulated.

SOLAR HEAT MULTIPLIER

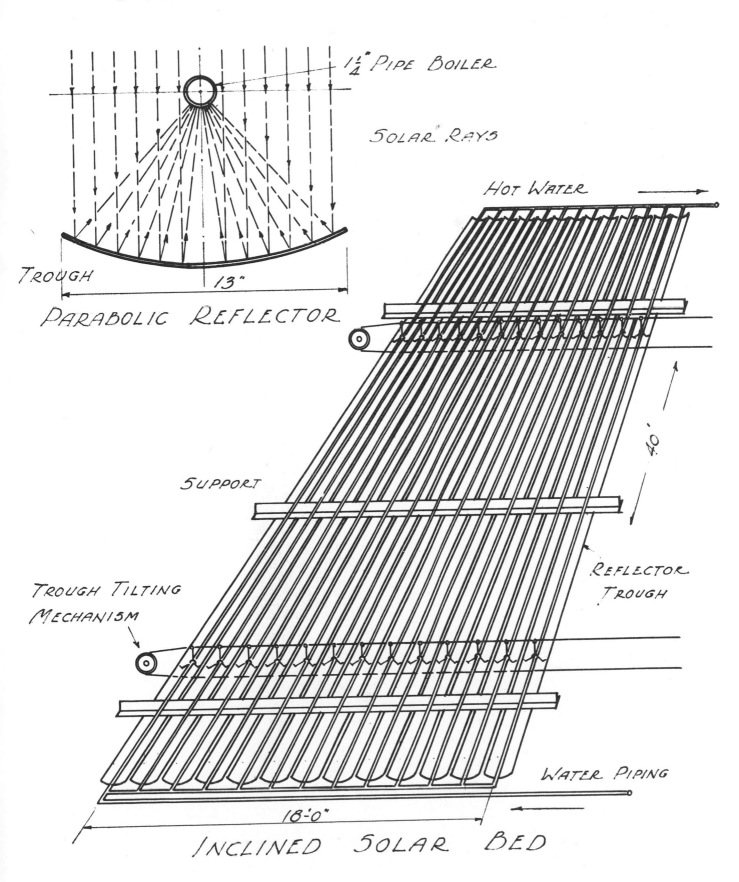

$1\frac{1}{4}$" PIPE BOILER

SOLAR RAYS

TROUGH

13"

PARABOLIC REFLECTOR

HOT WATER

SUPPORT

40'

REFLECTOR TROUGH

TROUGH TILTING MECHANISM

WATER PIPING

18'-0"

INCLINED SOLAR BED

CONICAL COLLECTOR—
FLUID BEARING

The radiant rays of solar heat are bountiful and constant. But they are mild in temperature and in strength when measured against the heat of combustion in the burning of conventional fuels while generating steam or electric power. This means that the reflecting or collecting area has to be large, involving considerable land use, expansive light construction, and safety in time of storm.

The essential features of a conical collector are several: the cone must rotate and elevate so as to be constantly in focus and normal to the sun; the structure must be little affected by the wind, with the reflecting surface bright and clear; and it must be lightweight and easy in operation. These all suggest electronic controls, fluid and/or roller bearings, an open cone with the vertex removed so the wind can pass through, and glass mirrors of high reflectivity.

The energy required to operate the mechanisms produces relatively little friction, due to the extremely slow motion, both in elevation and azimuth. However, it is imperative that friction be reduced as much as possible, since the energy available from the collector is relatively small in measure. To that end a floating bearing may be practicable. (A floating bearing is similar to an inverted cup or a diving bell in a tank of water, with entrapped air under automatically controlled pressure to support the load at the desired level.) The shape and position of the fulcrum actuating the cone must be determined mathematically to best suit the particular geographic location.

CONICAL COLLECTOR—FLUID BEARING

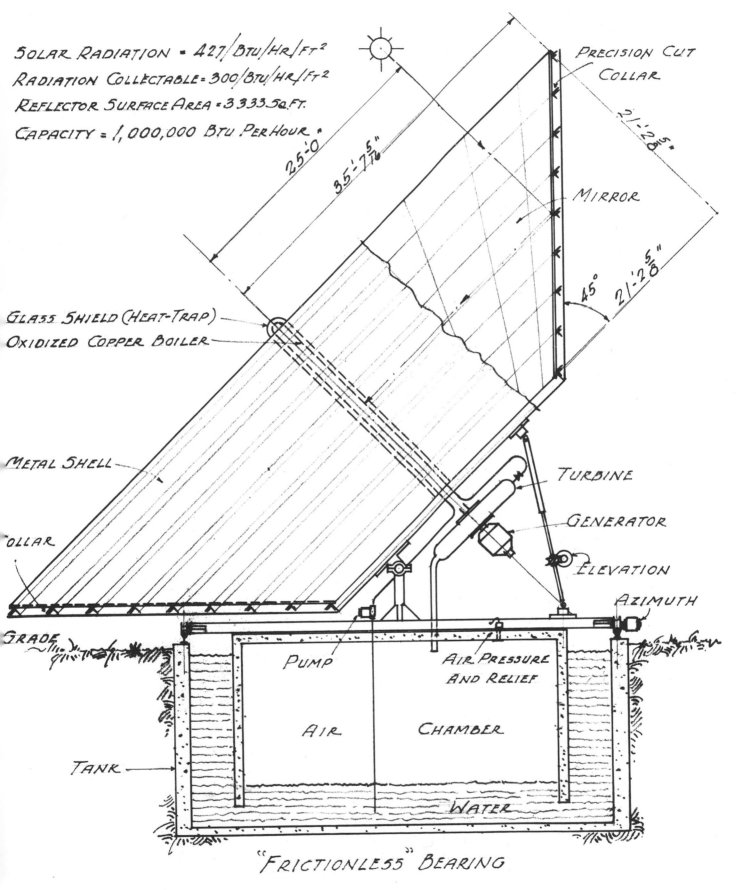

SOLAR RADIATION = 427/BTU/HR/FT²
RADIATION COLLECTABLE = 300/BTU/HR/FT²
REFLECTOR SURFACE AREA = 3333 SQ.FT.
CAPACITY = 1,000,000 BTU PER HOUR.

PRECISION CUT COLLAR

21'-2 5/8"

MIRROR

45°   21'-2 5/8"

25'-0"

35'-7 5/16"

GLASS SHIELD (HEAT-TRAP)
OXIDIZED COPPER BOILER

METAL SHELL

COLLAR

GRADE

TURBINE

GENERATOR

ELEVATION

AZIMUTH

PUMP

AIR PRESSURE AND RELIEF

AIR   CHAMBER

TANK

WATER

"FRICTIONLESS" BEARING

43

HEAT-COLLECTING
MECHANICAL SUNFLOWER

As the natural flower, with its cup-fashioned petals, turns toward the sun to obtain the maximum benefit, so can we derive an additional benefit by concentrating the sun's rays to generate steam for power. This requires the combining of reflected sunlight with the direct sunlight to raise the temperature up to and above the boiling point of water.

The hexagonal absorber or boiler shown in the drawing is principally the same in design as the conventional solar heater commonly used in Florida. This consists of a dull black absorbing surface for the face of the boiler, and a dead air-space of one inch or more located behind a sheet of window glass but in front of the boiler; thus creating what is known as a heat trap. The reflected heat from the six petals or lobes should equal about three times that of the direct sunlight at the absorber or boiler, for an intensity of 4:1.

Obviously the collected heat must be contained and this is done by insulation as with any boiler. Also the volume is kept low (1:1600) by retaining the water in its liquid state, though under pressure, and allowing it to flash into steam only in a steam chamber or turbine. With a reflector/boiler ratio of 3:1, a water temperature of 300°F should be easily attainable. That temperature and its corresponding pressure is quite appropriate for high pressure steam and standard iron pipes.

There is a choice between a conical collector and a polygonal figure, as shown, with flap-type reflectors. If only a few collectors were required the latter would get the preference because the mirrors are flat and easily available; but if there were many collectors needed, the conical collector with curved mirror reflectors would be the more efficient, desirable, and economical.

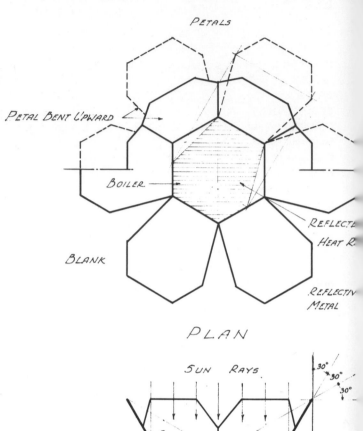

METALLIC SUN FLOWE

PETALS

PETAL BENT UPWARD

BOILER

REFLECTL
HEAT R.

BLANK

REFLECTIN
METAL

PLAN

SUN RAYS

30°
30°
30°

REFLECTIVE SURFACE
(INTERIOR)

STEAM

INSULATED BOILER

60°

$1\frac{3}{4}$ $\frac{7}{8}$

SECTIONAL ELEVATION

SOLAR HEATED WATER, STEAM, AND ELECTRIC POWER

To obtain the maximum amount of heat, the collecting surface must have the best possible heat absorptivity and it must be located where the sun is high and the air is clear. Under the most favorable conditions attainable on the surface of the earth, a temperature of 190°F is about the upper limit. This is just a little less than what is required to produce steam at atmospheric pressure, but woefully short of what it takes to make useful power in this age of electricity. Fortunately, it is within the scope of our knowledge to bend the rays of sunlight, and by concentrating them, to raise the effective temperature far above that required to boil water; hence we can produce steam and with it electric power from solar heat.

In all kinds of work there is one best way in which a particular job can be done; this also applies to the utilization of the heat and power, which arrives on earth continually from the sun, known as solar energy. Eventually, solar energy may be converted directly to electric energy, in strength that can be used for industrial power;

but for the time being it would be extremely gratifying if the heat of the sun were converted to steam and then to electricity. This would have a tremendous effect on the conservation of fuel and the minimizing of pollution.

The drawing suggests multiple service from stepped-up solar heat. The four sizes indicated can provide domestic hot water, also steam and hot water for heating and cooling, as well as part-time electric power. Another field that appears to be in the offing is that of pollution-free automobiles, using storage batteries for power.

The smaller-sized solar heat collector, as shown, could replace the typical hot frame type solar heater commonly seen in Florida. It could also charge batteries for automobiles, lawn mowers, and other mechanical tools. The second and third sizes could heat and cool homes, depending on latitude and climate, and the fourth and largest size could be used in series to generate electric power.

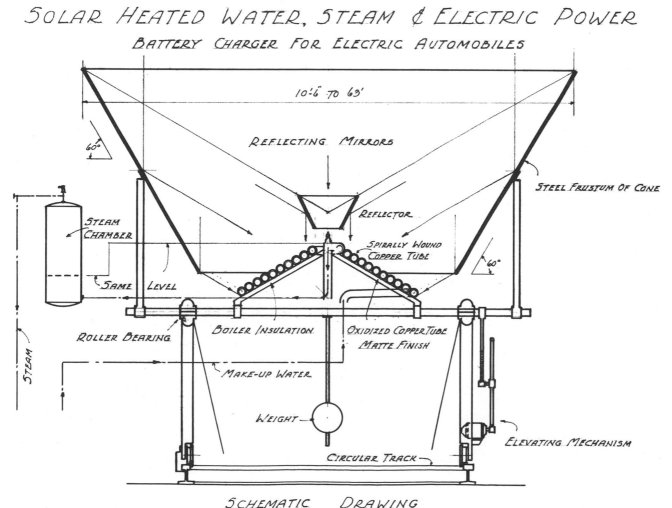

SOLAR HEATED WATER, STEAM & ELECTRIC POWER

BATTERY CHARGER FOR ELECTRIC AUTOMOBILES

SCHEMATIC DRAWING

THE HEAT-COLLECTING WALL

For many years solar heat had only limited applications—the greenhouse and the farmer's hot frame, for example. Out of these simple applications grew a useful principle that made solar energy practicable in a much wider variety of ways.

At first this use was embodied in the typical Florida solar heater, consisting of a glass-covered case about 10 cm deep, with serpentine copper tubes about 15 cm apart soldered to a backing sheet of black-painted copper. This was laid over fiberboard insulation and emplaced on the southern slope of the roof.

This type of installation had disadvantages. It was fixed; there was heat loss as the rays penetrated the glass; and a loss also in changing from one medium to another.

Then came a positive development. The sandwich-plate solar heater enables the heat to go almost directly into the water. The possibility of re-radiation is minimized by the low emissivity of the oxidized copper or aluminum of which the sandwich plate is made.

The panel may be laid on a roof as a low-cost conventional water heater. It may also be used as a heat-collecting exterior wall panel on the sunny side of a building; or placed anywhere that there is sufficient sun. The panels may be placed horizontally or tilted, fixed or pivoted.

SANDWICH PLATE SOLAR HEATER
ON
ROOF, EXTERIOR WALL AND IN YARD

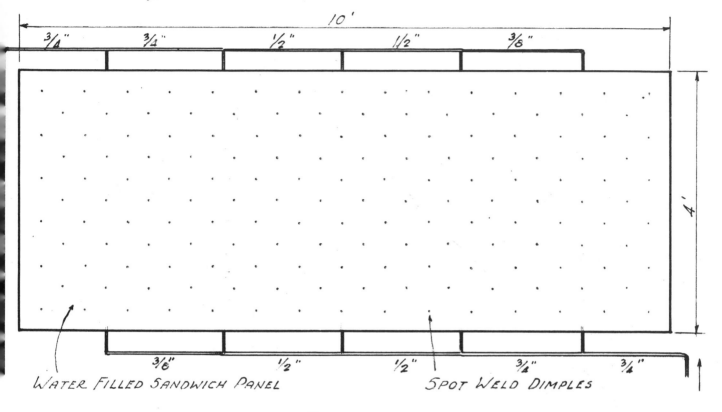

3/4"	3/4"	1/2"	1/2"	3/8"

10'

4'

WATER FILLED SANDWICH PANEL

| 3/8" | 1/2" | 1/2" | 3/4" | 3/4" |

SPOT WELD DIMPLES

PLAN

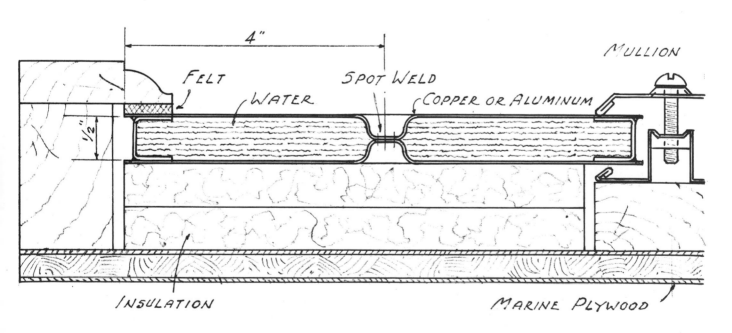

4"

FELT   WATER   SPOT WELD   COPPER OR ALUMINUM   MULLION

1/2"

INSULATION   MARINE PLYWOOD

DETAIL

SOLAR POWER PLANT

The feasibility of converting solar heat energy into electric energy depends on the cost of real estate, plant, and operation. It may be thought that because of the small amount of heat available per square foot of collector area, the cost of real property and plant might represent a handicap. However, this is offset and probably outweighed by the fact that the heat from the sun is free and there is no cost for fuel in the operation of the plant. Because sunshine is not dependable as a steady source of energy because of clouds and, of course, the darkness of night, some will argue that the cost of the plant would be higher than for conventional systems of power generation. This is problematical; except possibly as applied to the use of oil or natural gas for producing power.

Moreover, the benefits of invention and experience hold promise, and will favor the advent of solar power. Then too, the use of solar energy and power from the sea may one day become mandatory.

A solar power plant may be lightweight and the electronic controls can be simple; one control for elevation and another for azimuth. Resistance to wind should be at a minimum, and likewise any loss due to friction. This suggests a windbreak of evergreen trees and frustums of cones as collectors, the latter so the wind can pass through. It also suggests floating bearings to minimize the friction in turning and tiling the conical collectors while keeping them trained on the sun. The energy and mechanisms required for control can be reduced by arranging the collectors in multiple, thereby necessitating only one control for raising, and one for rotating, as for example, a solar power plant riding on a float. Actually it may develop that storage batteries and/or pumped water storage, together with an auxiliary source of power such as the jet engine, may be desirable or required.

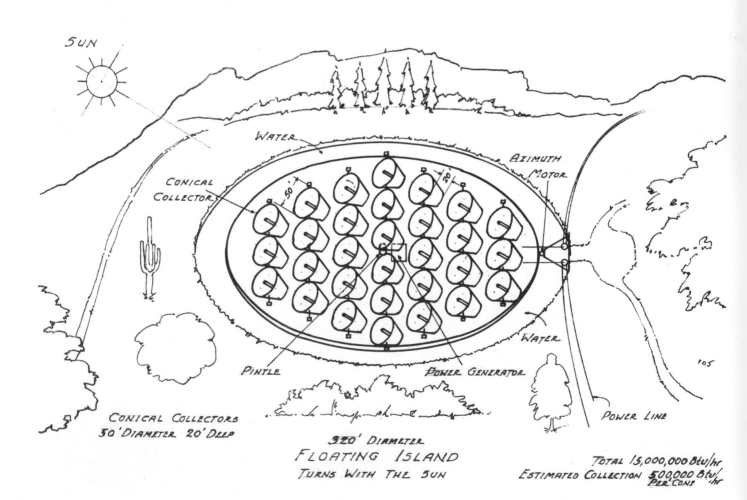

SOLAR POWER PLANT

CONICAL COLLECTORS
50' DIAMETER 20' DEEP

320' DIAMETER
FLOATING ISLAND
TURNS WITH THE SUN

TOTAL 15,000,000 Btu/hr
ESTIMATED COLLECTION 500,000 Btu/hr PER CONE

HEAT STORAGE

Heat collected by the conical collector or the long-range reflector/collector can be stored in tanks of water underground for heating when winter comes, and in tanks in a generating plant for immediate use in making steam or electric power. The tanks for power need heavy insulation; those for heating can be ordinary fuel oil tanks buried underground and encased in reinforced concrete which will withstand great pressure, ranging up to 300 pounds per square inch. Also, alkaline concrete is required to prevent the steel tanks from rusting. When the water temperature in the buried tanks is 200°F and above, the surrounding earth becomes powder dry and serves as a good insulator for retaining heat. The heat from the hot water in the tanks will penetrate only a short distance into the ground outside the concrete.

Most of the heat will be collected from March to November, when solar heat is plentiful. Collecting heat in fair weather in winter may be profitable, but it does not compare with the amount of heat that is available in summer.

Can heat be stored for so long? Yes. A study at the University of Connecticut proved that heat penetration into the earth is surprisingly low. A 4000°F thermite bomb can be rendered harmless by placing it in a pail of sand. Retarded action is what makes heat storage in the ground possible.

Uninterrupted low temperature heat is available to the heat pump due to an enlarging sphere of attraction, caused by moisture migration and improved conductance during the chilling cycle of the heat pump. Another important factor is this: measured by the sun, the height of the heating season is December 21; but actually, because of the residual summer heat in the ground, the midpoint of the heating season is delayed until February 1. What is more, the lowest temperature, at the five-foot level below ground, does not occur until March 1. By then spring is already having its affect in reducing the heating load.

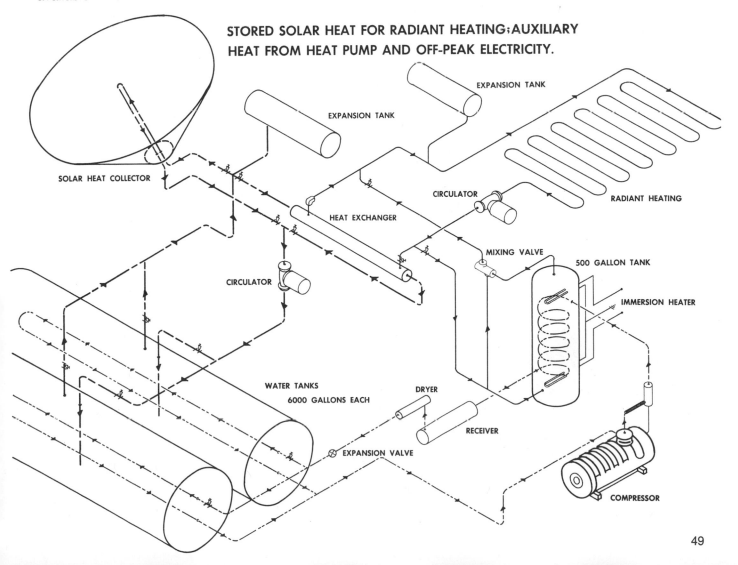

STORED SOLAR HEAT FOR RADIANT HEATING; AUXILIARY HEAT FROM HEAT PUMP AND OFF-PEAK ELECTRICITY.

HEATING AND COOLING WITH SOLAR HEAT

In many parts of the world the solar heat collector is as familiar a feature of the skyline as the holovision antenna. This collector makes it possible to heat buildings with hot water or steam in the winter and to refrigerate them in the summer. The part of the apparatus that we see, the conical collector, draws in the sun's rays. We don't see the equally important storage equipment.

More than fifty years ago some scientists foresaw that heating and cooling—in climates where the average temperature is 55°F or above—would be practical. Here is the description of such a system, written in 1980:

The sun's rays may be concentrated with a parabolic or a conical collector. For transferring the heat, a system of pipes containing high-temperature hot water or oil under pressure is probably the most suitable. Inasmuch as a large area of reflector surface is needed to collect the relatively small amount of available radiant energy, the collector presents an equally large surface to the force of the wind. This could become an important element in the cost of construction. From this standpoint, a hollow cone, sliced transversely into three parts and mounted in reverse order, will deliver the same amount of energy as a normal cone, and will offer a minimum resistance to the wind. This type of collector is easy to balance and to service. The reflecting surfaces may consist of mirrors or polished metals with a suitable finish. The boiler should be of copper, oxidized to a deep brown or black.

HEATING AND COOLING WITH SOLAR HEAT

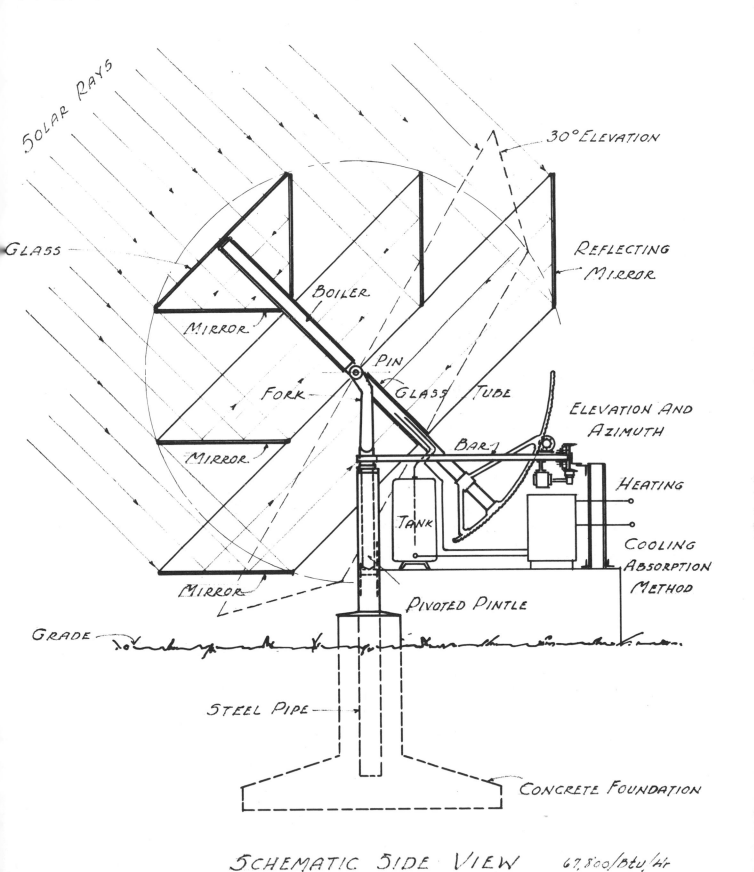

SOLAR RAYS

30° ELEVATION

GLASS

REFLECTING MIRROR

BOILER

MIRROR

PIN

FORK

GLASS TUBE

MIRROR.

BAR

ELEVATION AND AZIMUTH

HEATING

TANK

COOLING ABSORPTION METHOD

MIRROR

PIVOTED PINTLE

GRADE

STEEL PIPE

CONCRETE FOUNDATION

SCHEMATIC SIDE VIEW   67,800/Btu/hr

SOLAR HEAT FOR SEA WATER

Growing pollution of so-called fresh water lakes and streams led many people to seek new and reliable sources of water. Their search came back to a basic fact of our lives—the *ocean* is the source of our water; its interaction with the *sun* causes the rain that maintains our supply of fresh water.

This thinking led to the development of the solar still for seawater. Direct and reflected sunlight provides the heat. The target area or heat trap is lined with a metal of low emissivity like copper, but oxidized to a deep brown for absorption.

The boiler temperature will usually exceed 100°C. The condenser is positioned in the shade and painted with a terra cotta brown metallic oxide paint for maximum heat radiation and dissipation. The result is a continuous flow of distilled water.

SOLAR HEAT STILL for SEA WATER

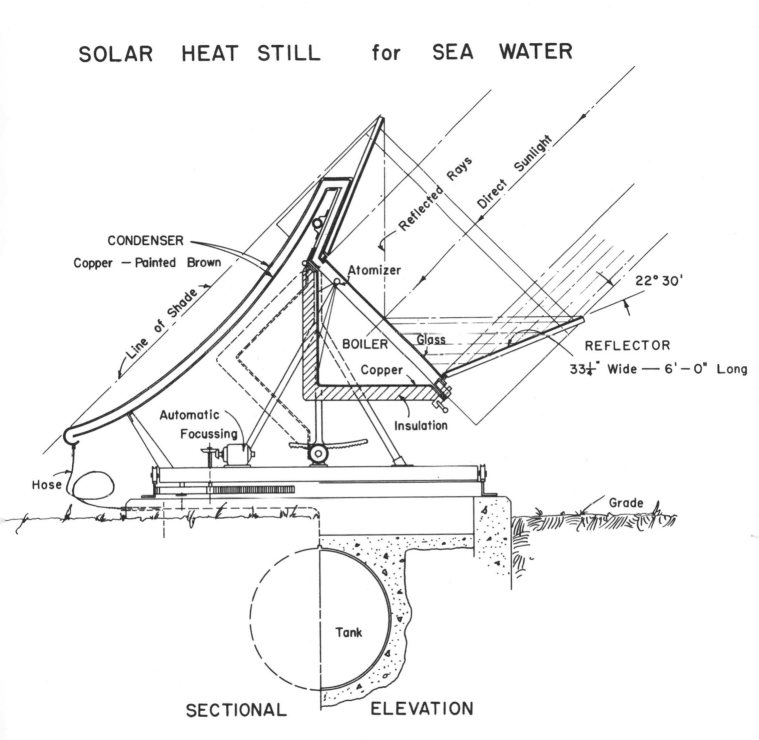

CONDENSER
Copper — Painted Brown

Line of Shade

Reflected Rays

Direct Sunlight

Atomizer

22° 30'

BOILER

Glass

Copper

REFLECTOR
33¼" Wide — 6'-0" Long

Insulation

Automatic
Focussing

Hose

Grade

Tank

SECTIONAL   ELEVATION

Water—the Blood of the Eart

The waterwheel furnished the first mechanical power in industry. Then it went out of fashion as other, more efficient power sources took over.

But now the waterwheel is back. This old device has been adapted to new uses; exploiting of energy from the tides, waves, and ocean currents. Of course today's waterwheel bears little resemblance to its ancestor. It floats on its side within a protective housing.

Here are contemporary comments indicating how this ancient implement had come to new life as a means of delivering clean, effective power:

In many cases the waterwheel is the simplest and most effective device for converting the energy of flowing water into electric power.

Floating power-units require secure and durable anchorage. Each project will present its own individual problems, but in general, deep water anchorage can consist of corrosion resisting cable, fastened to grillage of structural steel buried in dykes of broken rock, or the cables can possibly be weighted down with reinforced concrete "doughnuts." There must always be enough slack in the cables to allow surface vessels to pass in safety. There must also be a boom-type screen to prevent floating debris, including tree trunks and even boats from getting into the waterwheels or turbines. In relatively shallow water, piers, buttresses, and abutments, with struts instead of cables might be more suitable.

Horizontal waterwheels or turbines designed for buoyancy will make it easy on the bearings.

In the concept of the floating horizontal water-wheel, the float or hull in which the wheels are housed is shaped so the water of the current will converge and impinge on the paddles or vanes from a favorable attitude. The hull is intended to be of light, welded sheet steel, with flat deck and bottom. It will contain the necessary mechanical equipment, plus sufficient ballast to set the waterwheels at their proper depth. The generator is to be located in a weathertight cubicle, high and dry atop the deck.

The power units must be tethered, spaced, and held firm against all the forces of nature, along with the cumulative energy that is withdrawn from the flow. Their location may be lengthwise in the middle of the current, laterally from land to land, or built into a sluiceway. Anchorage may consist of concrete masonry, or of cables spanning the current either lengthwise or crosswise. The pivotal bearings of the wheels should be of bronze or stainless steel; being submerged, they will benefit from the natural cooling effect of the water. And, of course, the buoyancy of the air in the thick parts of the vanes will lighten the load on the bearings.

The hulls should be inexpensive, as there is no great stress, and the boxlike shape provides its own strength. The cables must be strong and durable, and may do double service by carrying not only the physical load but also power and signal lines. The entire construction is light in weight, is in plain sight, and the elements can be easily replaced or serviced. There can be an integral system of lights and fog horns for safer navigation, and as the draught is only around thirty feet, there is no interference with the normal habits of the fish. The more notable advantages are: a relatively low investment, a uniform and dependable source of energy and power, no fuel costs, and no pollution of the environment.

FLOATING HORIZONTAL WATER WHEEL

— PLAN VIEW —

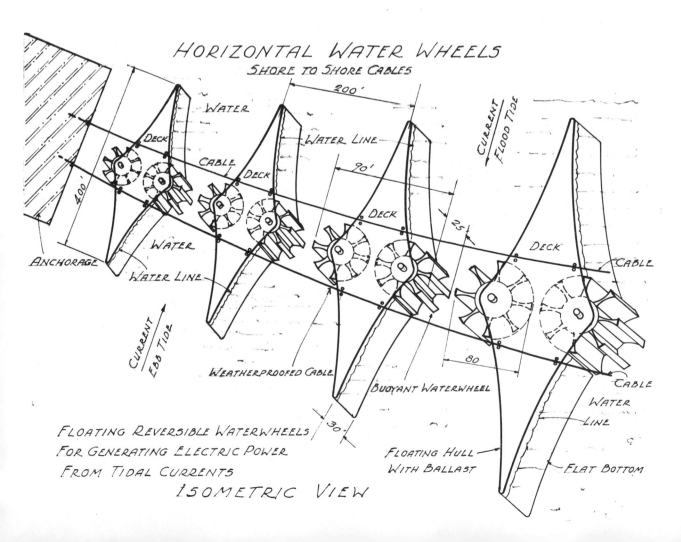

80'
400'
20'
WATER
WATERWHEEL 25' DEEP
(REVERSING)
VERTICAL FACE
ENCLOSURE
WATER
CABLE
CABLE
POLE
GENERATOR
DECK
HULL
HULL 30' DEEP — FLAT BOTTOM
WATER
CURRENT
EBB TIDE
BUOYANT WATERWHEEL
HELICAL SPRING
SPACING DEVICE
20'
20'
CURRENT
FLOOD TIDE
WATER
200'
ENCLOSURE WALL
DECK OF HULL

HORIZONTAL WATER WHEELS
SHORE TO SHORE CABLES

200'
WATER
WATER LINE
DECK
CABLE
DECK
90'
CURRENT FLOOD TIDE
400'
DECK
25
DECK
CABLE
ANCHORAGE
WATER
WATER LINE
80
BUOYANT WATERWHEEL
CABLE
WATER LINE
CURRENT EBB TIDE
WEATHERPROOFED CABLE
30
FLOATING HULL WITH BALLAST
FLAT BOTTOM

FLOATING REVERSIBLE WATERWHEELS
FOR GENERATING ELECTRIC POWER
FROM TIDAL CURRENTS

ISOMETRIC VIEW

UNDERWATER POWER PLANT

Of all the energy available from the sea, that from tides and currents is the more constant and reliable. The vertical rise and fall of the tides, although very powerful, is too slow and outside the realm of present day practicability. However, the energy in the currents that is produced by the tides is quite easily available under existing technology.

The simplest and oldest method is the use of a paddle wheel, and this is the most suitable where the current or flow is not fast. A more modern device is the underwater turbine, which is particularly effective where the current is relatively swift, as from three to six knots per hour. Regardless; in each and every case, the faster the current the better.

There are many localities along the coast, due to the natural formation of the land, where the tidal currents are quite favorable for generating electric power. There are many others where, by means of structural blockage and diversion, additional power can be made available. The prize locations in the East are in the Bay of Fundy and along the New England coast.

Probably the logical place to extract the energy from coastal currents is near the surface of the water, but in cold climates where ice floes and ice jams are to be expected, and shipping traffic is hazardous, something else has to be done. In the light of past experience with submarines, caissons, and tunnel work, an underwater power plant is not inconceivable. A scheme for hinged elongated buckets, fastened in parallel to endless chains and sprockets, supported and anchored at the bottom, and with suitable air locks in a power generating bubble will bear investigation.

In view of the tendency for flowing water to divert from, or shunt around an obstruction or resistance, only a small portion of the total energy of the currents can be claimed. Correspondingly the buckets and the construction in general can be light-weight. The efficiency depends largely on the force and velocity of the flow. This can be augmented by channeling, diverting, and funneling the water toward the buckets. The buckets would be broad for a gentle flow, and narrower in a swift current for the same output.

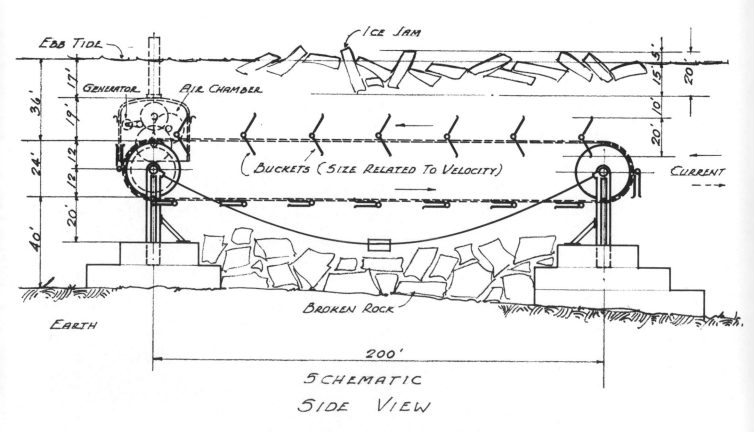

UNDERWATER POWER PLANT

SCHEMATIC SIDE VIEW

SUBMERGED WATER WHEEL
DOUBLE ACTION

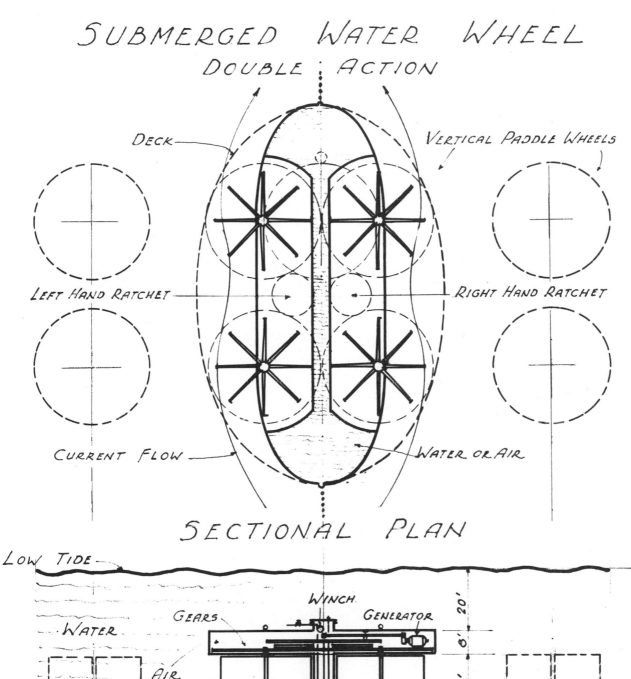

DECK

VERTICAL PADDLE WHEELS

LEFT HAND RATCHET

RIGHT HAND RATCHET

CURRENT FLOW

WATER or AIR

SECTIONAL PLAN

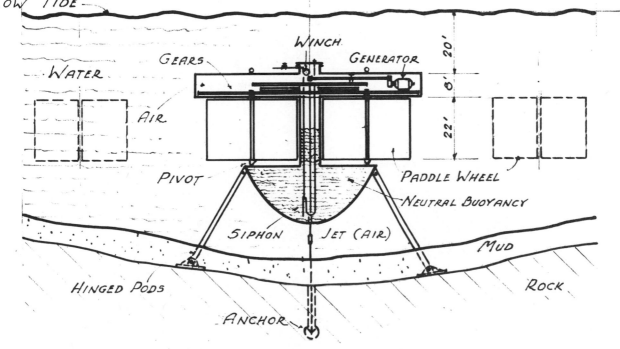

LOW TIDE

WATER

WINCH

GEARS

GENERATOR

AIR

20'

8'

22'

PIVOT

PADDLE WHEEL

NEUTRAL BUOYANCY

SIPHON

JET (AIR)

MUD

HINGED PODS

ROCK

ANCHOR

SECTIONAL ELEVATION

ENERGY FROM THE TIDES

Some forces in nature are extremely powerful, but too fast and dangerous to handle—lightning, volcanic eruptions, tornados.

By contrast, the motion of the tides is powerful but slow. However, the development of present-day mechanics makes it practicable to step up speed (though at the sacrifice of some power.) This has placed the tides within reach as a source of energy.

Obviously the total useful energy obtainable from the tides depends on their height. There is considerable unevenness in this power source, but on the other hand it works 24 hours a day. It is now economically feasible to take energy from tides two meters in height. Four meters and beyond is quite a good range.

One basic approach to drawing energy from the tides involves delayed-action weighted floats contained in a massive structure, strong enough to withstand the assaults of wind and sea. By locking and then releasing the floats on upstroke and downstroke high pressures can be attained. These working pressures are transmitted and put to work by water, oil, air or direct mechanical means.

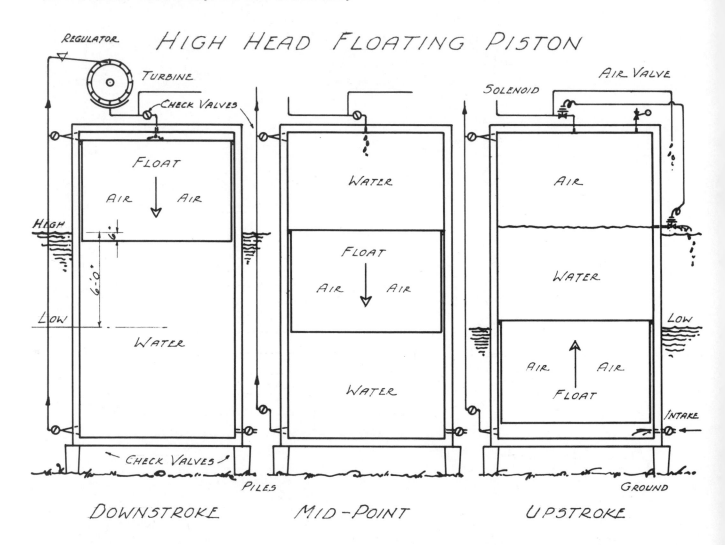

HIGH HEAD FLOATING PISTON

DOWNSTROKE   MID-POINT   UPSTROKE

TIDAL POWER.

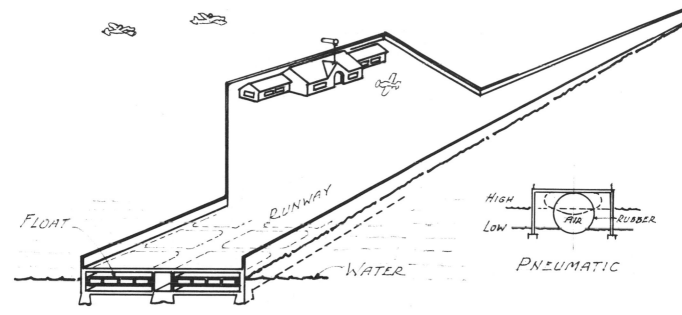

FLOAT   RUNWAY   WATER

HIGH   LOW   AIR   RUBBER

PNEUMATIC

AIR STRIP OVER WEIGHTED FLOATS

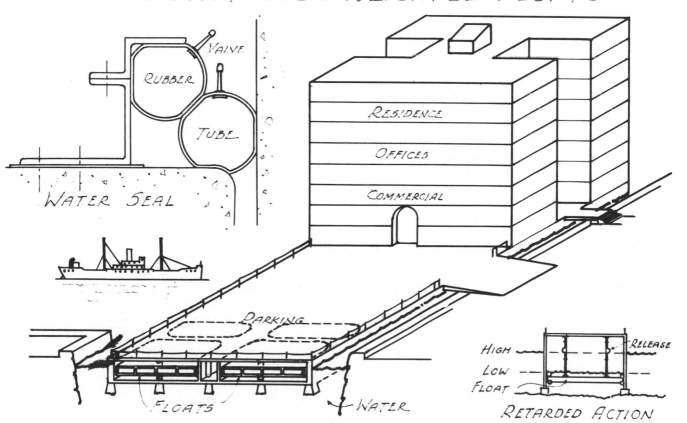

VALVE   RUBBER   TUBE

WATER SEAL

RESIDENCE

OFFICES

COMMERCIAL

PARKING

FLOATS   WATER

HIGH   LOW   FLOAT   RELEASE

RETARDED ACTION

DOUBLE ACTING POWER — COMMERCIAL

TIDAL POWER—THE GREAT CHALLENGE

For those who dreamed of turning the tides into power the great challenge was the Bay of Fundy. The rise and fall of water in this fluid wedge between Nova Scotia and New Brunswick ranks with the highest in the world, as high in some places as sixty feet.

Today this corner of the North American continent is served by power generated in the chain of stations along the bay. Looking back, we may see the challenge through the eyes of the planners of that time:

The power engineering challenge at the Bay of Fundy is very exciting. One day that great source of energy will be harnessed, and it will seem like discovering a new world. Present knowledge and techniques give us the means of trapping water at high tide, and generating power from its return to the level of low tide. To take additional power from the flow of water entering a storage basin at flood tide is equally possible. Moreover, the gleaning of mechanical power from the onrushing surface tidal water of the Bay of Fundy, with the use of waterwheels in buoyant frames is considered practicable, and less costly. Another possibility is to send sea water from the headwaters at high tide through penstocks to coastal power plants, as in Shediac or Baie Verte at the Northumberland Strait, or in the vicinity of Halifax, where the same high tide has a range of less than five feet.

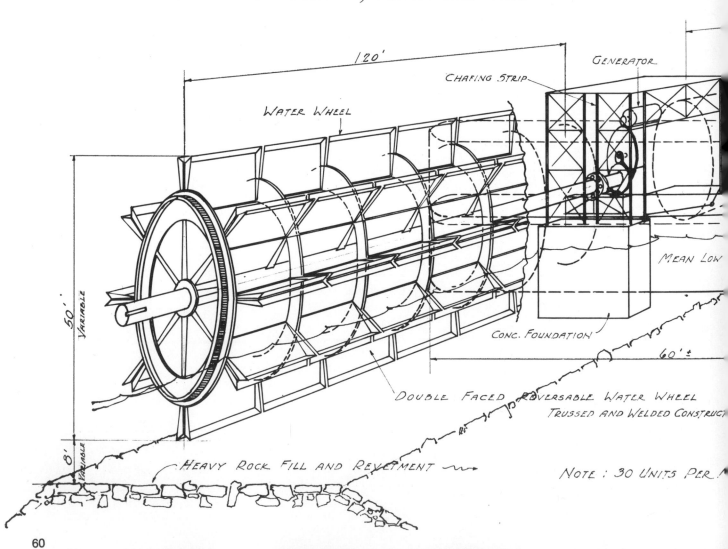

FOR BAY OF FUNDY
POWER FROM BUOYANCY, GRAVITY AND FLOW

120'

CHAFING STRIP

GENERATOR

WATER WHEEL

MEAN LOW

50' Variable

CONC. FOUNDATION

60' ±

DOUBLE FACED REVERSABLE WATER WHEEL
TRUSSED AND WELDED CONSTRUCT

8' Variable

HEAVY ROCK FILL AND REVETMENT

NOTE: 30 UNITS PER

With the Bay of Fundy installation as a model, other tidal projects were begun during the years 1985–2010. We are familiar with the large power conversion complexes at Cape Hatteras, Long Island Sound, and Cape Cod Bay. And many of us follow with great interest the new undertaking that has recently begun in the Bering Strait. These projects were foreseen many years ago. It is interesting to go back and examine the ways in which they were envisioned some seventy to eighty years ago.

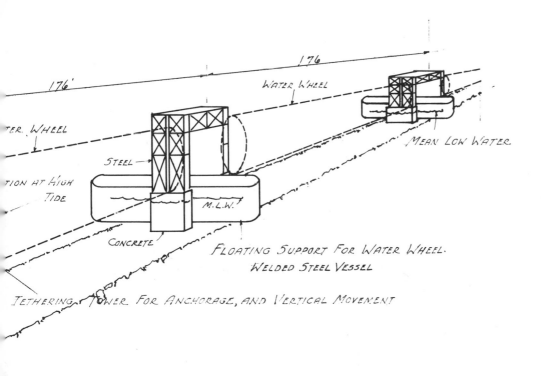

ELECTRIC POWER

FROM

TIDAL CURRENTS

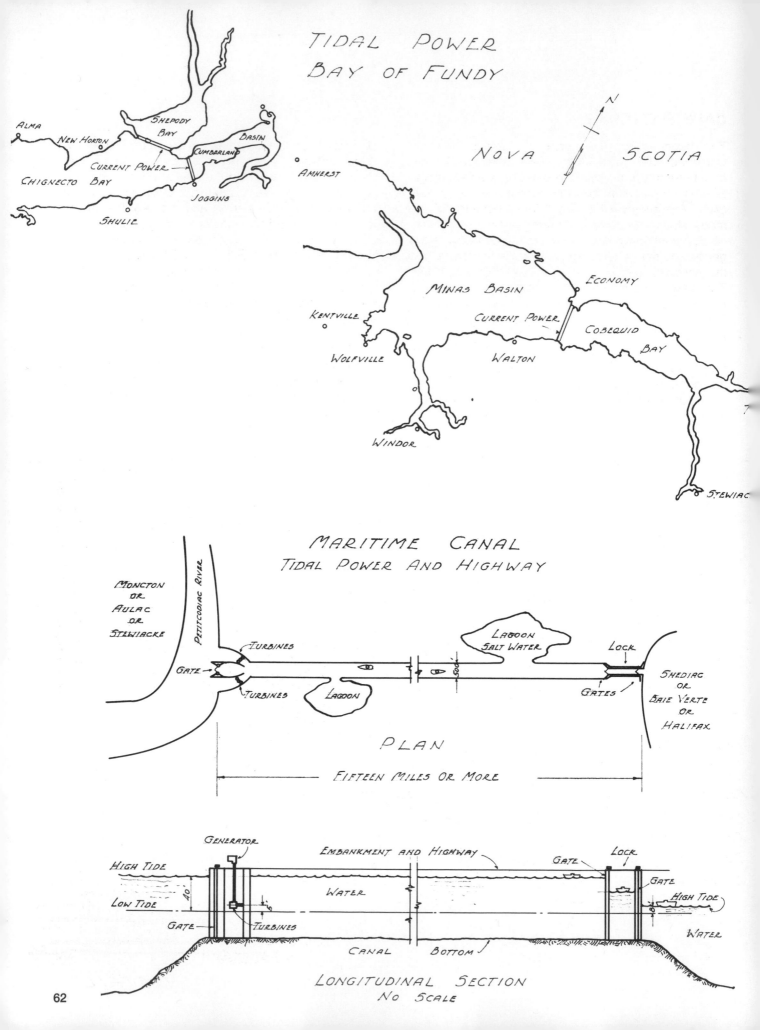

TIDAL POWER
BAY OF FUNDY

NOVA   SCOTIA

Alma
New Horton
Shepody Bay
Basin
Cumberland
Chignecto Bay
Current Power
Joggins
Shulie
Amherst

Minas Basin
Economy
Current Power
Cobequid Bay
Kentville
Wolfville
Walton
Windor
Stewiac

MARITIME CANAL
TIDAL POWER AND HIGHWAY

Moncton
or
Aulac
or
Stewiacke

Petitcodiac River

Turbines
Gate
Turbines
Lagoon

Lagoon
Salt Water

Lock
Gates

Shediac
or
Baie Verte
or
Halifax

PLAN
FIFTEEN MILES OR MORE

Generator
Embankment and Highway
Gate
Lock

High Tide
Gate
Low Tide
40
Turbines
Water
Canal   Bottom

Gate
Gate
High Tide
Water

LONGITUDINAL SECTION
NO SCALE

62

CAPE HATTERAS

The beaches on the coast of Virginia and the Carolinas are sandy, with reefs and shoals and treacherous winds and wave patterns. Cape Hatteras is known for its high seas and stormy weather, and is called the graveyard of the Atlantic because of the many ships that have foundered there.

The continental slope within the territorial waters is gentle and favorable, the currents are moderate, and the normal tide has a range of less than three feet. The waters are warm, and never plagued with ice. However, tropical hurricanes are a serious problem. Their normal course is northward along the east Florida coast, and they reach their greatest fury between Miami and Cape Hatteras, at which point they usually curve to the northeast and disappear into the northern Atlantic Ocean.

To obtain power from these waters involves structural, mechanical, and electrical engineering. The elongated paddle wheel has a place in the heavy surf, while bobbing generators are suitable in shallow deep water. It is important that the equipment and supporting structure be strong and in proper scale to best withstand the inclemency and vicissitudes of the weather. The paddle wheel should be built low and the vanes should be hinged in one direction so as to flap with a shortened lever arm (90°) when coming out of the water, and to fall, extending the lever arm as they come down the other side and enter into the water. Pivotal anchors should be built like a mule, with a thousand legs thrusting into the soft ocean bottom.

Probably the best way to install the bobbing generators is to cast the structural steel masts into a massive reinforced concrete footing. A system of jetties might also help. Much could be learned from a pilot installation.

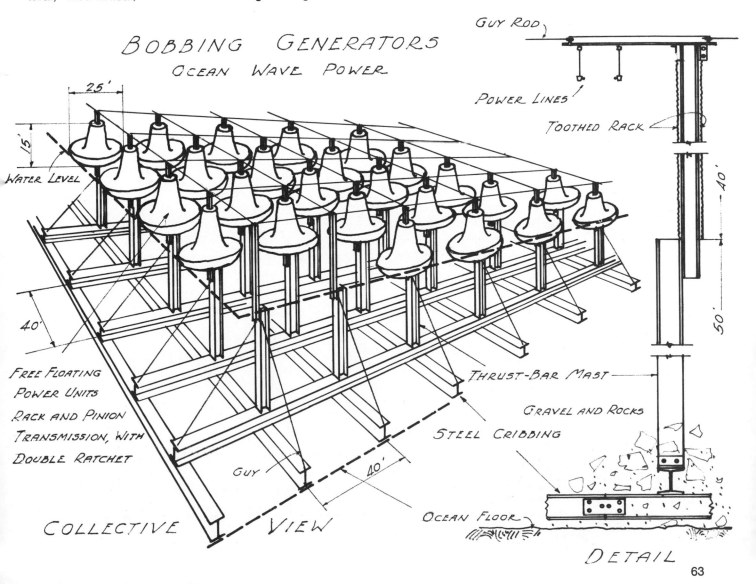

BOBBING GENERATORS
OCEAN WAVE POWER

25'

15'

WATER LEVEL

40'

FREE FLOATING
POWER UNITS
RACK AND PINION
TRANSMISSION, WITH
DOUBLE RATCHET

GUY

40'

COLLECTIVE VIEW

GUY ROD

POWER LINES

TOOTHED RACK

40'

50'

THRUST-BAR MAST

GRAVEL AND ROCKS

STEEL CRIBBING

OCEAN FLOOR

DETAIL

LONG ISLAND SOUND

Long Island Sound is almost a land locked body of water, about one hundred miles long and an average of around twenty miles wide. By partially closing the easterly end with a highway and power headwall an excellent source of power could be created. The water of the Sound flows in and out by virtue of the ebb and flow of the tide. With its tidal range of nearly six feet, the filling or emptying of six-foot depth of water every six hours results in a current velocity of about 6.75 feet per second at the westerly end or Hellgate Bridge. If the flow were narrowed in the same proportion at the easterly end, the volume and velocity would justify installing waterwheels and water turbines for the generating of power at both ends.

The new highway, the fishway, powerhouse, and bridges would be about twenty miles long; mostly through a moderate depth of water with a maximum depth of 150 feet. A bridge would span Plum Gut where the current is already very swift. By taking advantage of this natural container of active water, a sizable amount of electric power is available and a sorely needed short-cut to Rhode Island, Cape Cod, and maritime New England will accrue as a bonus. Reversible waterwheels and water turbines can be built into fixed positions with flanking jetties and guide walls to prevent interference by cross flow or counterflow.

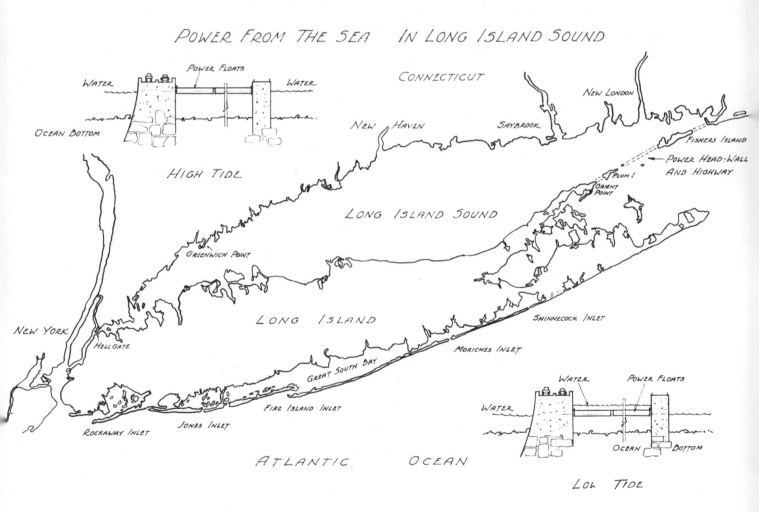

POWER FROM THE SEA   IN LONG ISLAND SOUND

POWER FROM BAY CURRENTS
REVERSING WATER WHEEL

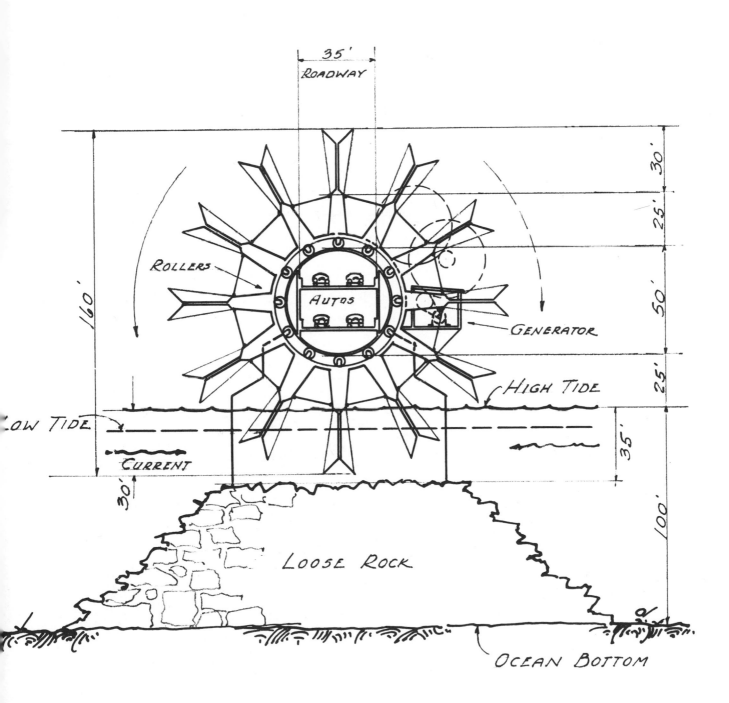

35'
ROADWAY

30'

25'

ROLLERS

50'

AUTOS

GENERATOR

160'

HIGH TIDE

25'

LOW TIDE

CURRENT

35'

30'

100'

LOOSE ROCK

OCEAN BOTTOM

SIDE VIEW OF WATER WHEEL

CAPE COD BAY AND CANAL

Because of the peculiarity of contiguous tidal currents influencing the flow in the area of Cape Cod Bay and Buzzards Bay, there is an inordinately high tide in Cape Cod Bay, and a strong current in the canal connecting the two bays. The tide on the north shore of Cape Cod has a range of 8.9 feet, and the two directional currents in the Cape Cod Canal has a speed of 2.4 to 4.5 knots, or an average velocity of around 6 feet per second. The Cape Cod Canal is entirely at sea level. It is about 8 miles long, 500 feet wide, and 28 feet deep. It is Government owned. The canal was built for coastal navigation and for military considerations. It is free to all shipping, and is under the supervision of the Army Corps of Engineers.

Cape Cod can be viewed as an undeveloped fairyland and the finest coastal summer resort area in the nation. It can also be seen as a place where an abundance of natural power is available, and improvements in the geography would be rewarding in the way of scientific beauty and easy access to the Cape. This suggests a highway and power headwall connecting Woods End (near Provincetown) and Monamet Point (near Plymouth). The value of such a highway is self-evident and the power that could be generated from the tide in Cape Cod Bay and from the current in the Cape Cod Canal would, in time, defray the cost. The depth of water in a slightly curved highway would not exceed 120 feet. The broken rock required for the deep water dyke is conveniently available from Government property on the mainland. Paddle wheels and submersible turbines could be considered.

POWER—IN CAPE COD CANAL

OPERATIONAL SCHEME

POWER—IN CAPE COD CANAL

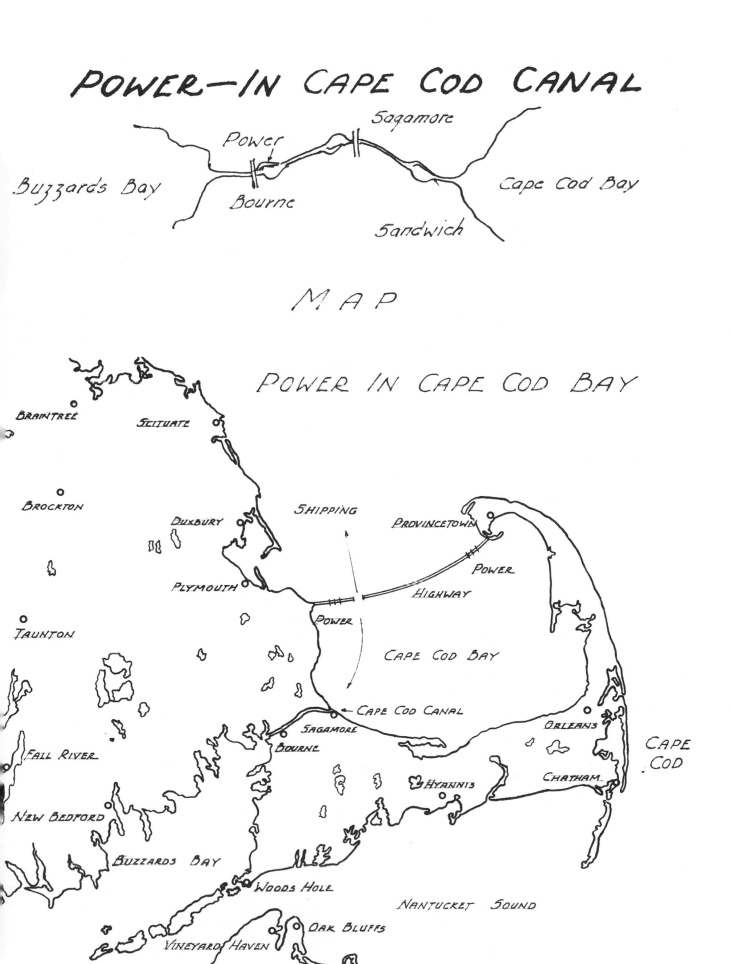

Power
Sagamore
Buzzards Bay
Bourne
Sandwich
Cape Cod Bay

M A P

POWER IN CAPE COD BAY

BRAINTREE
SCITUATE
BROCKTON
DUXBURY
SHIPPING
PROVINCETOWN
PLYMOUTH
POWER
HIGHWAY
POWER
TAUNTON
CAPE COD BAY
CAPE COD CANAL
SAGAMORE
ORLEANS
BOURNE
FALL RIVER
CAPE COD
HYANNIS
CHATHAM
NEW BEDFORD
BUZZARDS BAY
WOODS HOLE
NANTUCKET SOUND
OAK BLUFFS
VINEYARD HAVEN

BERING STRAIT—THE CLIMATIC THROTTLE

If the Bering Strait were closed except for a channel for ships and a sluiceway for generating electricity, it could have a far reaching effect on the climate and environment of countries whose shores border on the Pacific Ocean, north of the Equator. At present the Strait acts as a heat thief in passing the warmth of the Japan Current and Alaskan Current through to the frigid polar region. Moreover, it may also have a slight inductive influence on the flow of the Gulf Stream in the direction of the Arctic Ocean—all to no avail. Closing the Strait with an all purpose wall, or causeway about a mile wide, would be a simple matter. The land connection of Asia and North America at this location (once above water) is now covered with water about fifty six miles wide and sixty feet deep. The continental shelf in this area is of the widest on earth, indicating that here was as isthmus at one time.

The benefits of such a land fill would be many and varied. By raising the average annual surface temperature of the northern Pacific Ocean, the climate in southern Alaska, below the Arctic Circle, would be milder; the coastal area of Canada and the United States would be warmer and the water more suitable for bathing. The Asiatic shores would benefit likewise. This could be a boon to commercial and industrial development in Alaska and the Yukon, and could open traffic by rail or truck from across the United States or Canada, through to Asia and Europe without unloading. With an overpass at the Panama Canal and at the Suez Canal the continents of South America and Africa could be included, thereby a joint effort by the members of the United Nations would be justified.

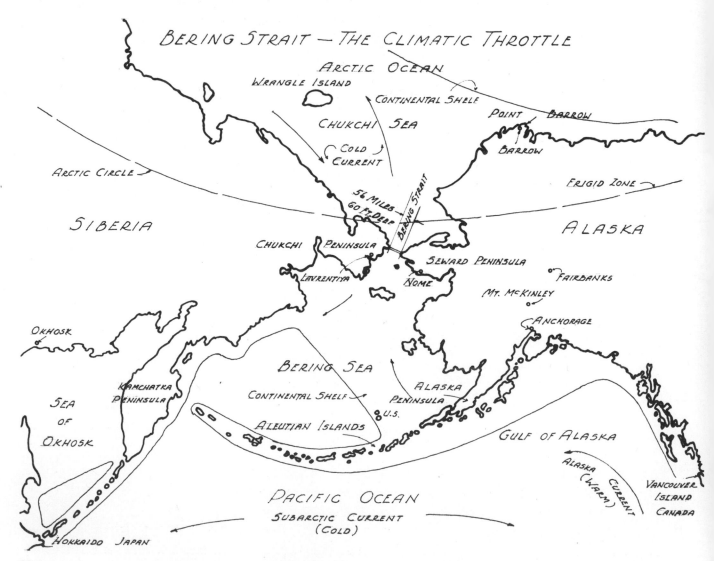

HYDRO-ELECTRIC POWER FROM THE GREAT LAKES

The great Canadian-American Niagara Hydro-Electric Project is still being planned and discussed. There are many problems. But we can be reasonably certain that by the latter third of the 21st century this project will be a reality providing clean power to a large area of two countries. Here is an example of the thinking that put this enterprise into motion three generations ago:

Since the energy shortage has become apparent, people are looking in all directions for possibilities of relief. Many have not realized the latent power in the water of the Great Lakes as it flows to the sea. By simple inquiry we find that four of these large bodies of water are on a relative plateau, about 600 feet above sea level. They do not drain a great deal of land but in themselves cover a very large area. The water flows collectively from lake to lake and eventually arrives at Niagara Falls in the Niagara River. The falls are ideal in size and height for the generating of

electric power. They are of course, also a thing of wonder and beauty.

To change or modify the physical nature of Niagara Falls requires courage and above all circumspection. The beauty of Niagara Falls must not be compromised, yet practicality and scientific progress in people's welfare deserve cautious consideration. Already a small portion (about 20%) of the water is diverted for electric power, by agreement between Canada and the United States. It seems that half the flow of the upper Niagara River could be used for power without losing much of the scenic attraction.

The sketch suggests additional apportioning of Great Lakes waters for power, on a 50-50 basis with Canada. To satisfy the environmentalists and for good common sense, it is thought that the additional take should occur only in the off-season, leaving the full flow and dramatic splendor during the days and nights of June, July, and August.

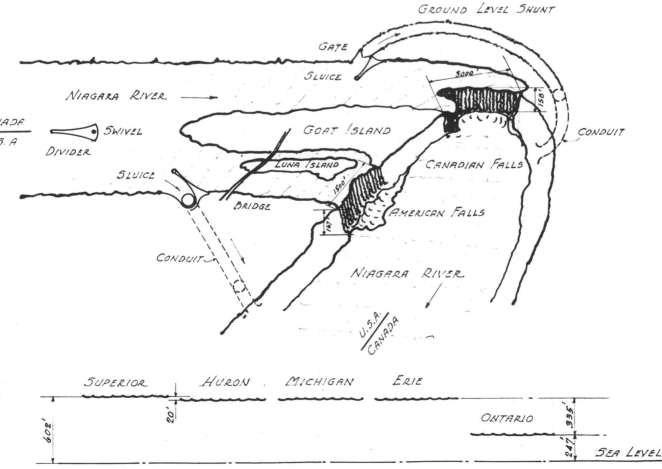

HYDRO-ELECTRIC POWER FROM GREAT LAKES

PROFILE OF GREAT LAKES

Power from the Waves

A bit of wood bobs on the surface of the water. It moves up and down as the waves pass beneath it, but seems always to remain in the same spot. This simple phenomenon has always been easy to observe. But not until the beginning of the 21st century was its message translated actively into a solution for the power needs of mankind.

We watch an object floating. The waves move shoreward, one after the other. The object does not ride into shore with the waves—although it may inch toward the land. Most of us know, in a general way, why this happens. The wave is not a moving mass of water. It is a force which lifts the surface of the water as it passes along. Each particle of water moves with the wave in an eclipse, but after the wave passes the particle comes back to just about the same place it started. The wave has force but it has no weight.

Sea waves, whether caused by wind, seismic disturbance, or the tides, can travel great distances while retaining their power. Because the water stays in essentially the same place as the waves pass along, there is little loss through friction. In the deep ocean a wave will role, unhampered, for thousands of miles. Vessels in the ocean constantly rise and fall on the surface, describing a simple orbital motion. If the vessel is stopped in the water it will not move very far from the point at which its forward motion ceased. It will ride to the top of the wave, roll to the trough, then begin to ascend the next wave. Ocean waves may achieve a height of more than sixty feet. (In the 20th century there were constant reports of waves as high as 120 feet, but these reports appear to have been enhanced considerably by the imaginings of the mariners of that time.)

Nowadays not many of us have a chance to observe deep ocean waves from the surface of the sea. There are no more "ocean liners"—passenger ships that once served as a principal means of intercontinental commercial transportation and later offered vacation tours. Today we fly over the water at altitudes of more than 65,000 feet and see little but a faintly wrinkled surface.

Our day-to-day observation of ocean waves is likely to be restricted to the shore. And here waves perform differently. Here is a description of shore waves, written around 1950:

As the waves approach the shore they reach water shallower than half their wavelength. Here their velocity is controlled by the depth of the water. The shallow bottom greatly modifies the waves. First, it refracts them, that is, it bends the wave fronts to approximately the shape of the underwater contours. Second, when the water becomes critically shallow, the waves break. Even the most casual observer soon notices the process of refraction. He sees that the larger waves always come in nearly parallel to the shoreline, even though a little way out at sea they seem to be approaching at an angle. This is the result of wave refraction, and it has considerable geological importance because its effect is to distribute wave energy in such a way as to straighten coastlines. Near a headland the part of the wave front that reaches shallow water first is slowed down, and the parts of it in relatively deep water continue to move rapidly. The wave thus bends to converge on the headland from all sides.

Another segment of the same swell will enter an embayment and the wave front will become elongated so that the height of the waves at any point along the shore is correspondingly low. This is why bays make quiet anchorages and exposed promontories are subject to wave battering and erosion—all by the same waves.

The final transformation of normal swell by shoal or shallow water into a breaker is an exciting step. The waves have been shortened and steepened in the final approach because the bottom has squeezed the circular orbital motion of the particles into a tilted ellipse; the particle velocity in the crest increases and the waves peak up as they rush landward. Finally the front of the crest is unsupported and it collapses into the trough. The wave has broken and the orbits exist no more. The result is surf. If the water continues to get shallower, the broken wave becomes a foam line, a turbulent mass of aerated water. However, if the broken wave passes into deep water, as it does after breaking on a bar, it can form again with a lesser height that represents a loss of energy. Sometimes the air trapped by the collapsing wave is compressed and explodes with a great roar in a geyser of water. On the other hand, if the bottom slope is long and gentle, as at Waikiki Beach in Hawaii, the crest forms a spilling breaker, a line of foam that tumbles down the front of the partly broken wave as it continues to move shoreward.

ROLLING ENERGY FROM SURF AND ROUGH WATER
— RESONANCE —

FEATHERWEIGHT FLOATS
SKIMMER TYPE

GUIDED HEFTY POWER ROLLERS

PRESSURE LOBE

WEIGHTED BALL

FOR SURF

80'

40'

OCEAN WAVE

15'

FOR ROUGH WATER

HEAVING FLOAT

10'

FLYWHEEL AND RATCHET TRANSMISSION
MAGNETO TYPE GENERATOR

300' PLUS

50'

WEIGHTED ROLLERS

CHAIN DRIVE

10 SECONDS

FLOAT IN ROUGH WATER

ANCHOR

For thousands of years the power of the surf was considered totally destructive. The surging and pounding of ocean waves broke down the shoreline. When winds became high the waves rolled inland to destroy nearby dwellings.

Jetties and breakwaters were built to interrupt the rolling thrust of the waves and diminish the erosive and destructive effect of the water as it reached the shallows. For a few, the surf became a medium of recreation. Surfboard riders sought out areas such as the northeast coast of Oahu where the beach was long and the surf consistently high. Bathers at beaches around the world enjoyed swimming and coasting on breaking waves.

But the beaches themselves were highly vulnerable to wave action. Ocean waves possess the power to wash away a mass of sand, leaving only mud. The erection of jetties offered some help in protecting sandy stretches of shoreline, but erosion remained a perplexing problem.

Around the middle of the last century various students of the waves such as Willard Bascom, began to formulate ideas leading toward the neutralization of the destructive effect of surf and the exploitation of wave power for positive effects. Their thinking involved the planned emplacement of breakwaters to shape and control the incoming waves, and the development of devices to translate wave energy into usable power.

We may still see the first fruits of these pioneers in the "paddle wheel" installations in such places as Bar Harbor, Maine, and Seal Rock, California. As the following contemporary comments demonstrate, the paddle wheel—even then—was not considered the definitive answer.

DEEP-WATER OCEAN WAVES

WAVE LIFTING EFFECT ON BOATS

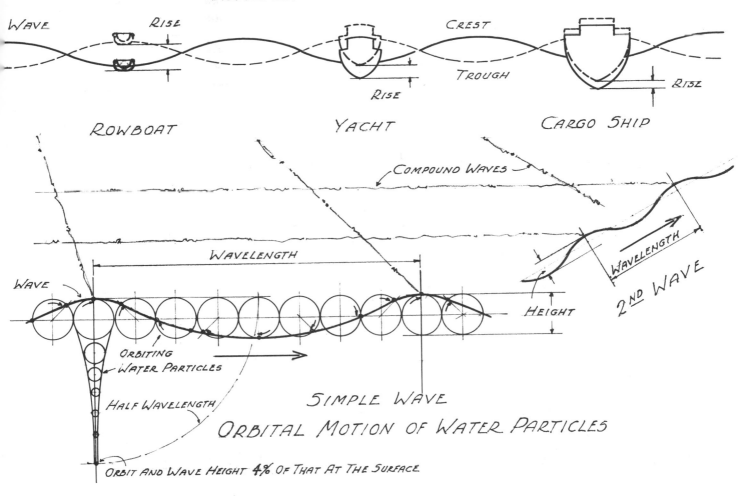

WAVE   RISE   CREST

TROUGH   RISE

RISE

ROWBOAT   YACHT   CARGO SHIP

COMPOUND WAVES

WAVELENGTH

WAVELENGTH

2ND WAVE

WAVE

HEIGHT

ORBITING
WATER PARTICLES

HALF WAVELENGTH

SIMPLE WAVE

ORBITAL MOTION OF WATER PARTICLES

ORBIT AND WAVE HEIGHT 4% OF THAT AT THE SURFACE

The paddle wheel method of utilizing the energy of the surf is probably the simplest, though far from the most attractive. The operation embraces the simple mechanics which are presently in vogue, somewhat resembling the mechanism of the windmill, but including a flywheel and governor to smooth out the spasmodic bursts of energy that occur at about ten-second intervals. These systems have a place, as interesting novelties, for generating power on an island, or for private shore property. A bank of storage batteries might be a desirable adjunct.

The choice of design depends on the peculiarities of the region or locality, in respect to tide, climate, and exposure. The energy yield will also depend on whether the installation is located on a point of land or in a bay or in a cove; the point or promontory may have as much as a two to one advantage. If there is the desired surf one must expect considerable wind, in which case the cost of structure may have some influence in determining the size and breadth of the paddle wheel.

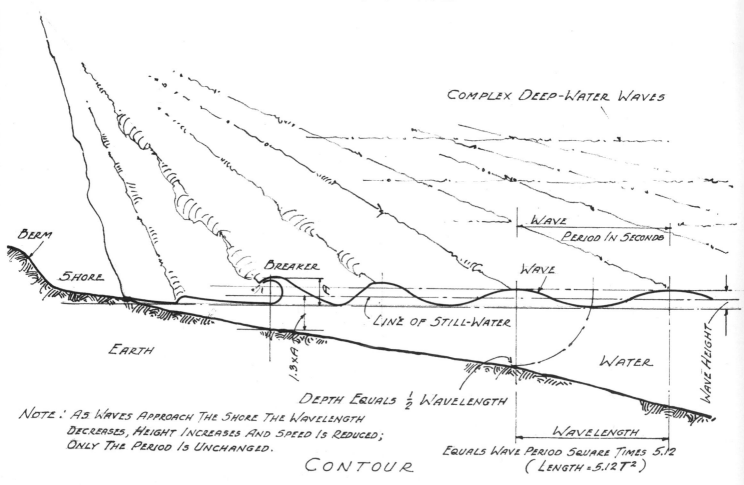

CHARACTERISTICS OF THE SURF
SHALLOW-WATER WAVES

COMPLEX DEEP-WATER WAVES

BERM

SHORE

BREAKER

EARTH

$1.3 \times A$

LINE OF STILL-WATER

DEPTH EQUALS $\frac{1}{2}$ WAVELENGTH

WAVE

PERIOD IN SECONDS

WAVE

WATER

WAVE HEIGHT

NOTE: AS WAVES APPROACH THE SHORE THE WAVELENGTH DECREASES, HEIGHT INCREASES AND SPEED IS REDUCED; ONLY THE PERIOD IS UNCHANGED.

CONTOUR

WAVELENGTH

EQUALS WAVE PERIOD SQUARE TIMES 5.12
(LENGTH = $5.12 T^2$)

OCEAN WAVES—WHIPLASH— HYDRAULIC RAM

There are four kinds of waves, namely, deep sea, shallow water, breakers, and tidal flow. Those of the deep sea may be complex, due to an intermingling of waves from various directions. Those in shallow water, with a depth of less than half the wavelength, have more or less a rolling action in the direction of the shore. Then there are the breakers in which the wavelength progressively diminishes as the wave approaches the beach or land; these have a horizontal component. Tidal flow is usually accompanied by waves that are also directional, generally toward the shore.

The waves that are best suited for energy exploitation are those in water 50 to 300 feet deep, usually classified as shallow water waves. These waves may act quite like those in mid-ocean with erratic rocking, winding, and up-and-down motion, which is conductive to the collecting of energy with mechanical contrivances designed for controlling and utilizing this wild action. How this can best be accomplished is still unknown, but a significant beginning has been made in the design of the familiar bell buoy.

Another idea involves a concentrated blow into a high pressure pump as shown, compressing either liquid or gas. The blow may be caused by one of several means. The one illustrated makes use of the weight, fluidity, and relative incompressibility of water, as a ram or train of force, with the rushing water in a pipe resulting in what is known as "water hammer." The heavy balls on the masts are intended to produce a whiplash effect and to augment the dipping and tossing of the float, the better to amplify the dashing of the constrained water against the plungers of the pumps.

The energy that can be captured from any one wave in this manner may appear infinitesimal, but when it is multiplied by the number of waves that pass per day (6 × 60 × 24) it will build up to an impressive amount. As a part-time source of power this device can contribute much to prolonging the availability of power from burning fuel.

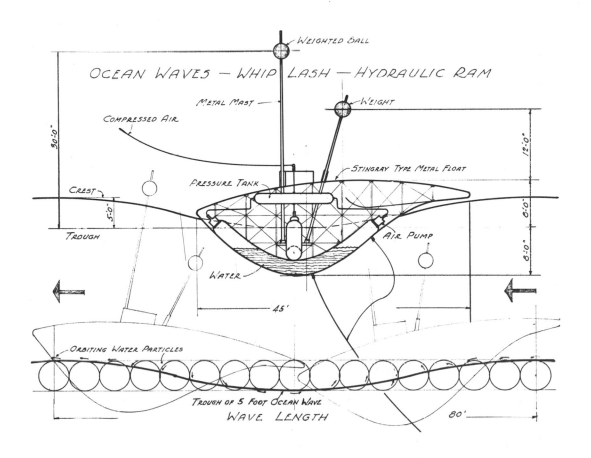

DOUBLE-ACTION OCEAN WAVE PUMP

The most striking characteristic of surface wave motion is the lack of frictional resistance. Since the water through which a wave travels goes practically nowhere except up and down, its power can be exploited easily.

A wave six feet high and a properly designed float can actuate a ram to deliver water hundreds of feet high into a storage reservoir, or to an accumulator at a lower level for direct hydraulic power.

An early and basic use of this principle consisted of an anchor, a tubular shaft containing a pump, and an airtight hollow metal float. The float sinks during the upstroke of the pump and rides high during the downstroke. The resulting resistance energizes the pump.

The higher the wave and the smaller the piston, the greater is the effect. Distance to point of delivery matters little, because of the very low frictional loss.

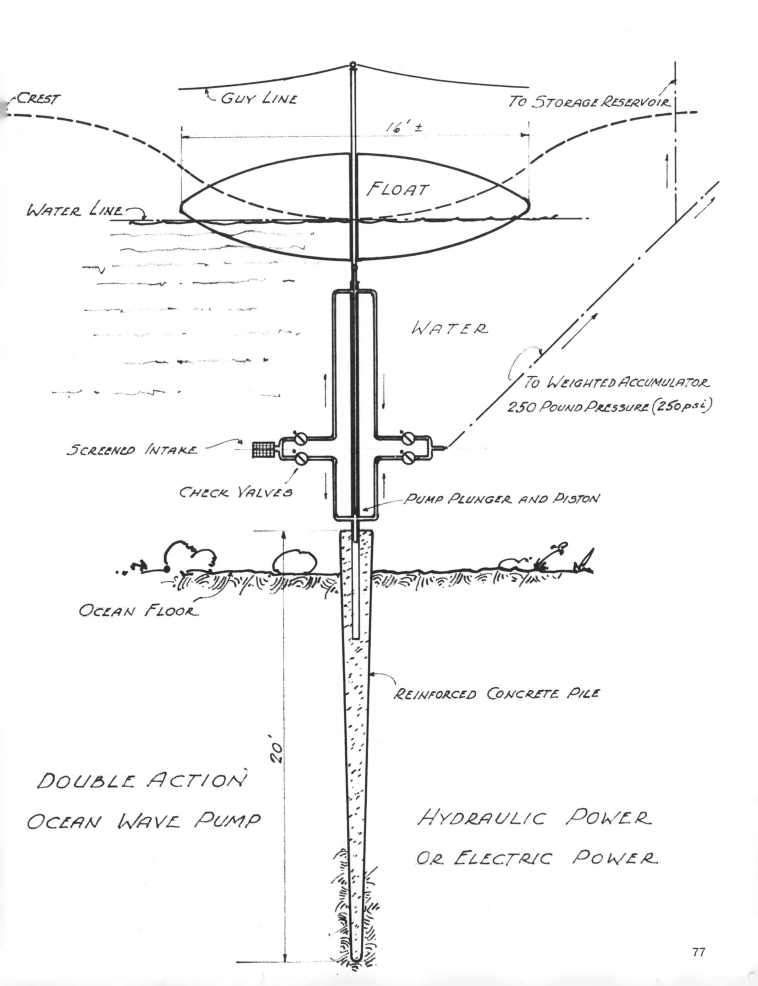

CREST   GUY LINE   TO STORAGE RESERVOIR

16' ±

FLOAT

WATER LINE

WATER

TO WEIGHTED ACCUMULATOR
250 POUND PRESSURE (250 psi)

SCREENED INTAKE

CHECK VALVES

PUMP PLUNGER AND PISTON

OCEAN FLOOR

REINFORCED CONCRETE PILE

20'

DOUBLE ACTION
OCEAN WAVE PUMP

HYDRAULIC POWER
OR ELECTRIC POWER

WIND AND WAVE POWER— RESONATOR

Deep water waves differ from the shallow water ones in that they are a jumble of waves that cross one another, with the result that they recklessly toss about in no particular pattern, but still with have the characteristic up-and-down motion of the water as a wave passes through. The energy in a simple wave is easy to determine, but that of complex, deep water waves can have different values under various conditions.

As wind and waves move along together, it is possible to take power from both the horizontal flow of the wind and the vertical swaying motion of the waves, to obtain total power. One drawing shows a buoyant, inverted, bell-shaped float with a free-swinging, weighted pendulum and eight air pumps in a ring which are used to transmit the whiplash energy of the waves into the float in the form of compressed air. The windmill on top, by using other mechanical means, also contributes to the store of compressed air, possibly for delivery to an adjacent power generating float for conversion to electric power.

The second shows a resonating float of similar form, but with a different mechanism. In this case a free-swinging, suspended, weighted iron ring strikes an iron ring, which is connected to the pistons of a radial air compressor. The compressed air is stored in the billowy walls of the float, and in turn is used to generate electric power in a turbine and generator designed for the purpose. Whether the electricity is generated inside the float or at a central point, and whether the wind is also used for power, depends on the location and local conditions. In either scheme, the function of the anchor is only to prevent drifting, while the materials and structures must be capable of absorbing a great deal of stress, strain, and abuse.

OCEAN WAVE RESONANCE FOR POWER

SCHEMATIC

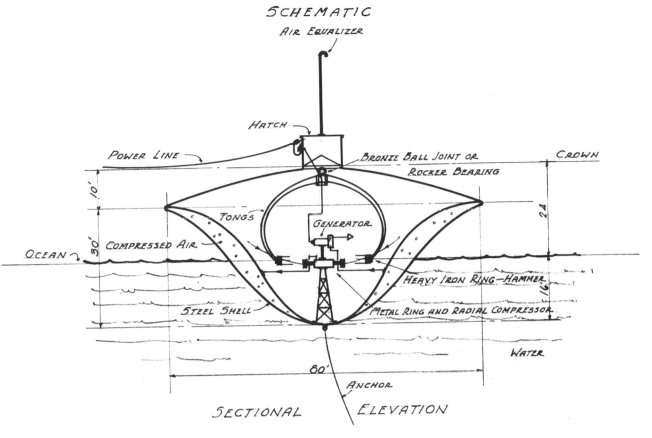

Air Equalizer

Hatch

Power Line

Bronze Ball Joint or Rocker Bearing

CROWN

Tongs

Generator

10'

30'

Ocean

Compressed Air

24

Heavy Iron Ring—Hammer

16'

Steel Shell

Metal Ring and Radial Compressor

Water

80'

Anchor

SECTIONAL   ELEVATION

WIND AND WAVE POWER

⋅⊹ RESONATOR ⊹⋅

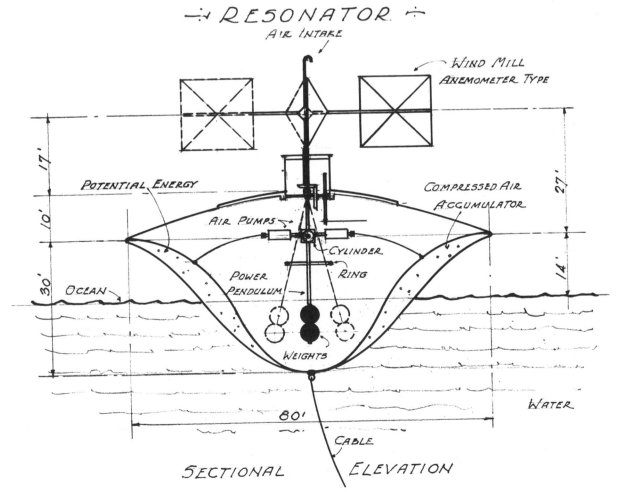

Air Intake

WIND MILL ANEMOMETER TYPE

Potential Energy

Compressed Air Accumulator

17'

27'

Air Pumps

10'

Cylinder

14'

Power Pendulum

Ring

30'

Ocean

Weights

Water

80'

Cable

SECTIONAL   ELEVATION

WAVE AND LEVER POWER IN COASTAL WATERS

Obtaining power from ocean waves can be accomplished in many ways, and as with all things there is one best way for any particular set of conditions. The choices become fewer in deep water outside the continental shelf, at depths of one hundred fathoms or more, where the wavelength remains constant for great distances. Quite different is the behavior of waves in shallow water, over the continental shelf, in depths of one hundred fathoms and less. Here the wavelengths shorten and the heights lessen as the waves approach the shore. This is, of course, influenced by the contour of the ocean bottom and a myriad of natural conditions contiguous to and peculiar to a certain location.

Consider an average location with a water depth of about forty to eighty feet, and with a history of waves that have not exceeded forty feet in height. The drawing is intended for such conditions and locations, of which there are many. Specifically it assumes a storm wave 20 feet high and an average wavelength of 150 feet, in water around 60 feet deep. There are three floats, positioned in an equilateral triangle to obtain maximum exposure to the waves, and to obtain a power stroke every half wavelength (within about five seconds). The floats are hemispherical, thus limiting the upward thrust to an amount equal to the weight of twenty feet of water, with the limit controls allowing for a safety factor of two. The available power is enhanced by the lever arm from the buoy to the hinged fulcrum. This can result in an additional advantage, amounting to a ratio of two-to-one or better.

Through this scheme the wave energy can be converted to mechanical or electric power by several means. One way: the energy captured from the up-and-down motion of the floats is transmitted to a system of gears or high pressure pumps, producing mechanical power or electricity for whatever use it can be put to, including the production of hydrogen.

WAVE AND LEVER POWER IN COASTAL WATERS

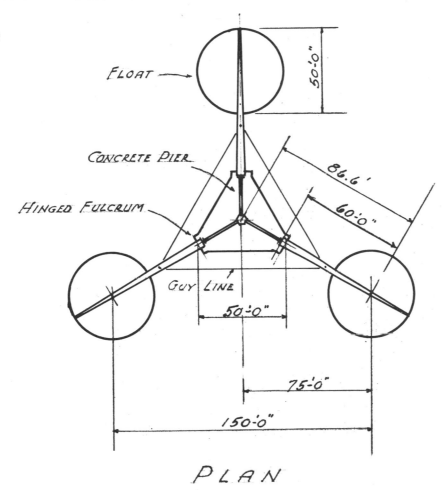

FLOAT

CONCRETE PIER

HINGED FULCRUM

GUY LINE

50'-0"

86.6'

60'-0"

50'-0"

75'-0"

150'-0"

PLAN

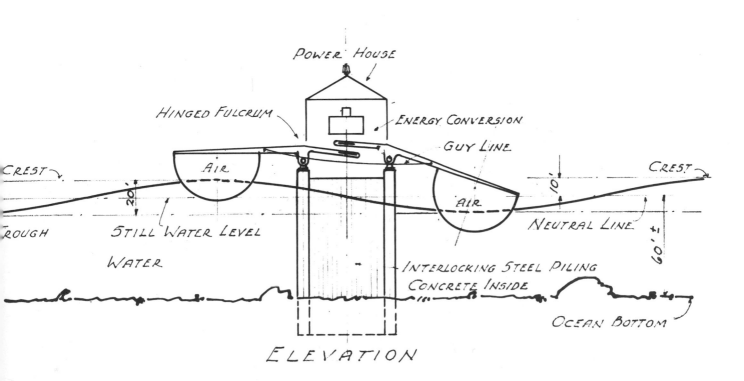

POWER HOUSE

HINGED FULCRUM

ENERGY CONVERSION

GUY LINE

CREST

AIR

CREST

AIR

10'

TROUGH

20'

STILL WATER LEVEL

WATER

NEUTRAL LINE

60'±

INTERLOCKING STEEL PILING

CONCRETE INSIDE

OCEAN BOTTOM

ELEVATION

SURF AND BREAKER POWER

The idea of harnessing the power of the surf caught on in the last century, and sturdier and more sophisticated devices were designed and tested. Some succeeded, others failed; but even the failures helped designers to improve ways of implementing the principle.

Today's student may be interested in the comments of a contemporary observer, as contained in the following excerpt:

By nature a small amount of energy can create a relatively large wave or surf. And whether a wave is wind blown or is a smooth surface swell, its motion or effective force is more up-and-down than horizontal. Of course, there is considerable forward thrust in the falling water at the breakers as the waves approach the shore. The available power may not be as great as it might seem, yet this together with the possibility of saving a beach from both damage during violent storms, and from the sometimes mysterious shifting of sands, warrants the building of a power unit and jetty in some localities.

There are several possibilities for extracting power from the surf. The elongated waterwheel of a design appropriate to the height of surf appears to be the most suitable. Because of the devastating forces of wind and storm, the wheel and structure must be firm, limber, and strong. These requirements are a challenge to new and original thought. A waterwheel of conventional design could well result in considerable weight, which in turn might impose a heavy load and produce excessive friction at the bearings, and thus consume a large part of the energy contained in the waves.

SURF POWER—SLINKY SHAFTING—LIQUID BEARINGS

WELDED STEEL SHAFTING
AND REVOLVING FLOATS

10'

MEAN HIGH WATER

SHORE LINE

AIRTIGHT FLOAT
AND SHAFTING

SEA WATER

75'

TANK

WATER

75'

GENERATOR

JETTY

ACTION OF NATURAL WAVE

ACTION OF WAVE IN TAILRACE.

PADDLE

SHORE PROTECTION AND SURF POWER

GUARD

BUMPER   WELD

AIR

WATER

AIR

LONG SECTION AT FLOAT

OCEAN   WAVES

WAVE

STILL WATER LEVEL   HEIGHT

WAVELENGTH

The problem of design is that of dealing with intermittent energy and impact, so as to keep within low limits of frictional resistance at the bearings, and to accept the untoward forces of wind. The wave height and frequency will vary, ranging from two to six feet for design conditions, with periods or frequency of five to ten seconds. The wavelengths may be thirty to one hundred feet. This calls for light but rugged construction; analysis suggests a buoyant shaft and floating flywheels in tanks of water, whereby making the friction in the bearings or floats almost nil, considering the extremely low coefficient of friction between the metal surface of the floats and the water, at so slow a speed.

The simple waterwheel soon gave way to the concept of the "power bucket." This device involved a tilting bucket, with geared transmission of power, connected with a ratchet clutch. The mechanism was coupled with either an electric generator or a pump.

One of the early bucket-type installations was described this way:

The scoop-type bucket rolls with the force of the water, and returns to the starting position after its power stroke. The return takes place, with the aid of an overbalancing weight and escape slots in the bucket, during the time interval or period between waves, ranging from about five to ten seconds—more nearly the latter if the waves and breakers are high. It is assumed that this device will work best in locations where and when the waves are three or more feet in height, and occur often enough to justify the cost of installation. The mechanical principle involved is somewhat analogous to a watch with its powerful but slow-moving mainspring and its train of gears for stepping up the speed to a practical rate.

SURF ACTION

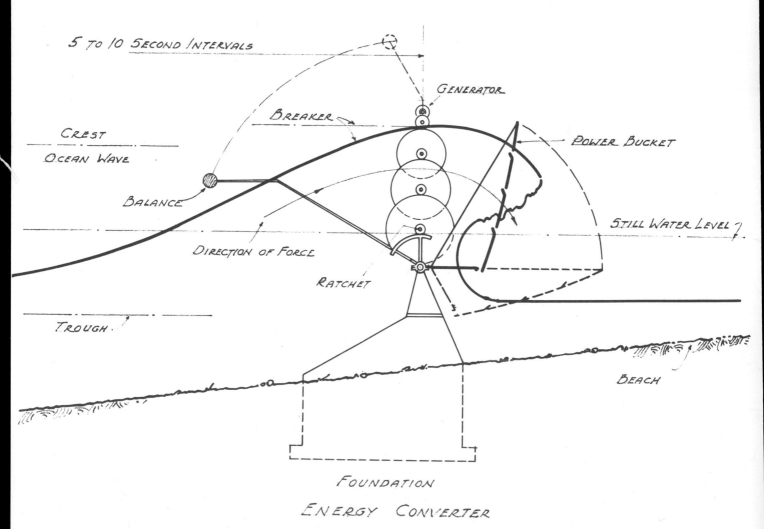

ENERGY CONVERTER

Paddle wheel and bucket-type installations are still operating satisfactorily in many places along the coasts of the world. But an increasing amount of the power we derive from onshore waves comes from simpler devices which are geared to the lifting motion of the waves. The principle is related to the simple example of the bit of wood bobbing on top of the water as the waves pass along. This up-and-down activity has been harnessed in such places as Yaquina Head on the windswept Oregon coast.

At first, the wave-lift arrangements were small-scale, designed to enable individual householders to derive power from the nearby surf. Builders in the early years of the 21st century were guided by such commentaries as this one:

As we know now, there are many locations where the force of the ocean waves can be made useful. For a simple, small-scale installation a "weightless" float with attendant cables, sheaves, and weighted work tower is probably the most practicable and suitable. By these means the vertical motion of the waves produces a vertical movement of a sliding weight in the work tower, which results in a gravitational force that can actuate a pump for water or air, or possibly actuate a mechanism based on the principle of the hydraulic ram.

The practical details as to size and kind of equipment can best be learned from experience with a trial installation, and from inquiry into the various methods of obtaining motive power and putting it to advantage in a particular way. It is assumed that an empty oil tank, six feet in diameter and forty feet long, riding on four to six foot waves would deliver enough power to run a moderate-sized onshore workshop. It could also provide electricity for the home and for beach and boat dock lighting.

WAVE LIFT TO DROP HAMMER

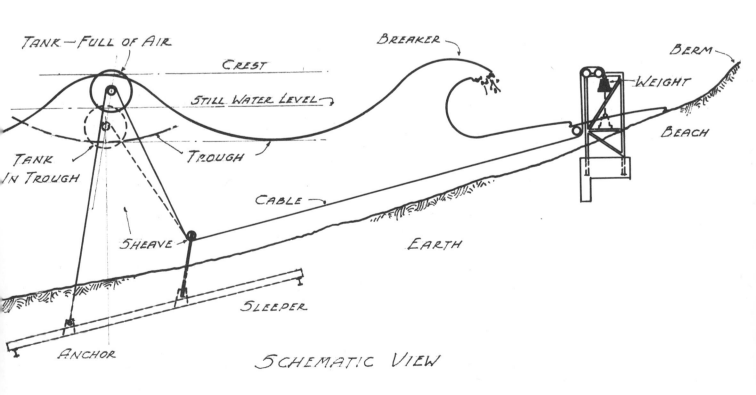

SCHEMATIC VIEW

There was a severe limitation on the use of natural surf as a source of power. Too few coastal areas embodied the right combination of elements to produce the kinds of waves that could be exploited for this purpose. So, since the natural coastal configurations were not numerous enough, it became apparant that *artificial* means would have to be found to direct and concentrate the energy of the breaking waves and convert it into positive use.

Now, of course, there are numerous such plants, at Cherbourg on the French coast and at Wick in Scotland, to name just two. The essence of the method is to impose on the shoreline man-made structures that simulate the action of a sea wall, trap the energy of the artifically-broken wave, and then turn it into usable power. In this way we came to derive great benefit from waves in areas where there were no natural configuration that would produce the

desired effect. Some of the earliest planning for an artificial sea wall is quoted here:

When the common wave gets close to shore and becomes a breaker (due to drag at the bottom and backwash from the shore), the molecules describe an elliptical orbit and become part of a horizontal thrust as the breaker spills the water on the land. Since the breaker cuts off the reaction or balancing half of a normal wave, trough to trough, a strong horizontal force in addition to the vertical force, occurs, as at the face of a cliff or sea wall. Together they are amplified to the maximum by designing a device to collect the force of the rising wave and that of the wave rushing in and dashing against the ceiling of a dug-out pocket, thus simulating the action on a rugged coast line with no beach.

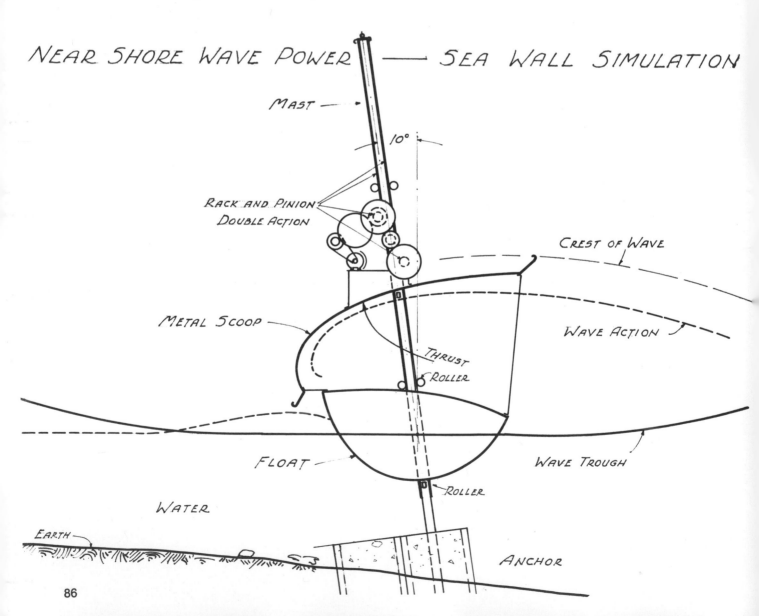

NEAR SHORE WAVE POWER — SEA WALL SIMULATION

The large coastal power plants with which we are familiar today operate using concepts that have grown out of these earlier developments. Here is how the first big dual-power converters worked:

Wave energy near the shore can be separated into forces acting vertically and horizontally by employing the principle of the seesaw and the hydraulic ram. As the ocean waves approach the shore in relatively shallow water, the wavelength shortens from say 500 feet to about 100 feet. Though the wavelength and height will vary, the frequency remains more or less constant at around ten seconds.

One basic means of obtaining dual power involves a circular tank, half-full of water, its length half that of the design wavelength, and its position athwart the waves. At both ends of the tank there are hermetically sealed tanks filled with air. These tanks can be of any practical design length, or probably the same as the first tank—half that of the design wavelength.

The longitudinal tank functions as a hydro-pneumatic ram. As the water in the tank rushes from one end to the other during each wave, the pistons send compressed air into a high pressure receiving tank, from which it activates a rotor and develops electric power in the generator. Coincidentally, the end floats or bounding buoys convert the uplifting force of the waves into mechnical and electric power by means of gears.

DUAL POWER FROM ONSHORE WAVES

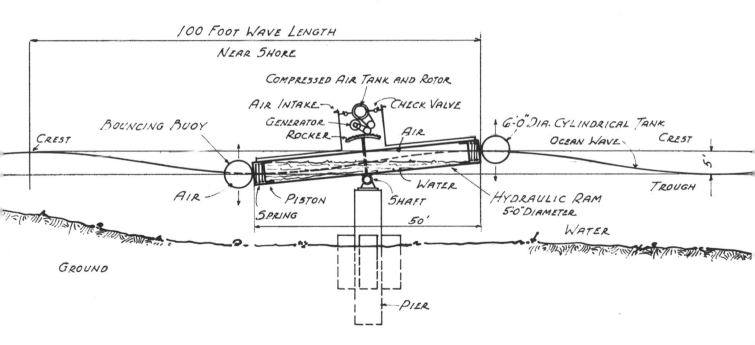

SCHEMATIC CROSS SECTION

More Energy from the Oceans

During the initial experimentations in the last century, onshore wave converters were found to be effective—but their scope was limited. Installations designed to use the power of the surf could be built only in places where the configuration of the shore line was favorable.

So in the final decades of the last century, planners—seeking hungrily for sources of clean power—lengthened their sights. They began to look outward, beyond the breakers, to the waves rolling along the surfacee of the deep ocean. The total power of all the waves in deep water is almost unimaginable, approaching infinity. As long ago as 1970 certain scientists concluded that their technology was adequate to the task of making a small percentage of that vast power available.

At first the thinking centered on the emplacement of fixed ocean platforms, similar to the offshore oil rigs that dotted some coasts during that era. But the difficulties in making such platforms strong enough soon became apparent. The following is from a document of the time:

A pilot installation was tested in 1982. It consisted of an anchor, a tubular shaft containing a pump, and an airtight hollow metal float. The energy for operating the pump, comes from its resistance to the rising and falling of the float. If the float were only to ride the waves at its normal water level there would be no power; but if the float were sunk to its full height during the upstroke, and then rode high during the downstroke (because of work done by the pump in each case), a considerable amount of useful energy could be realized.

We have also found that the length of the piston and the size and shape of the float must be designed to suit the average conditions indigenous to each locale. The higher the wave and the smaller the piston, the greater the effect. The distance from the wave to the point of delivery matters little in so far as friction loss is concerned, because of a low velocity; however, an appropriate and practical size of pipe line is desirable.

In a violent sea almost any kind of mooring may fatigue and fail. It seems, therefore, that the machinery for extracting power should ride the waves, resisting them only enough to skim off a measured amount of the energy, all within the safe structural stability of the design installation.

The motion of the water due to the waves is both perpendicular and circular. The water within each wavelength moves up and down, backward and forward, and yet it goes nowhere except for a slight amount of drift influenced by the movement of wind and sea. The drift is slow and has little force but it may be constant and must be dealt with.

Since the energy in ocean waves is mainly in the up-and-down motion, the apparatus for converting that energy to power must be designed with resistance or reaction to vertical forces. For that purpose a bobbing generator seems to be indicated. Such a system tends to nullify inertial resistance, as when a swift acting force meets the slow acting resistance of inertia. This can be illustrated by a rifle bullet passing through a free-standing board without causing the board to fly or even to quiver. Another example is in the breaking of a brick with the edge of ones hand, in a karate chop.

The drawings show an idea for obtaining power from the waves in deep water. The set-up consists of a floating spar-type mast with a large stabilizing plate at the bottom, a bobbing float, racks and pinions with clutch and ratchet feature, stepped-up transmission

gears and flywheels, electric generator, and weather-proof housing.

In operation, the pinion gears run up and down the mast, as the float on which all gears are mounted rides the waves. The clutch and ratchet in each of the pinions is positioned so that one takes power in the upstroke and the other in the downstroke. The mast contains air for buoyancy and has a broad plate at the bottom to resist sudden surges, upward or downward.

Theoretically, a wave can come along—do its work of raising the float with somewhat of a jerk—before the mast and stabilizing plate have had time to respond. The force will be nullified by the downward thrust of the float as it descends into the trough of the next wave. This, of course, depends on the mast being long enough to reach into the relatively quiet water below.

The power of the ocean wave was finally captured with the greatest efficiency by the double action wave pump. Since the water through which a deep-sea wave passes goes practically nowhere except up and down, a properly designed float can exploit this constant motion with little frictional loss.

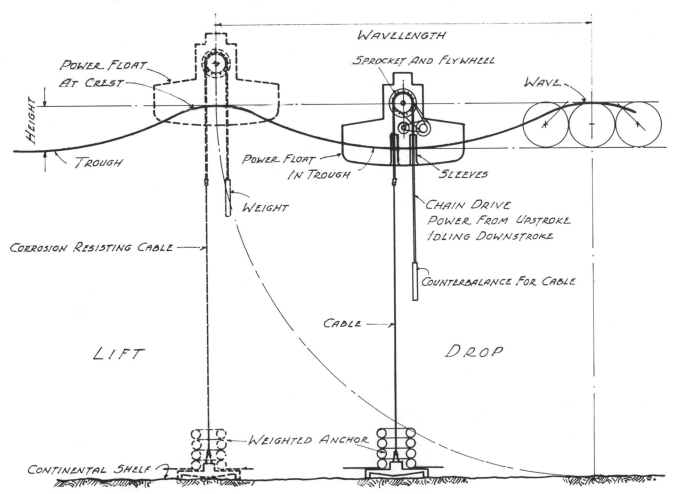

WAVE POWER IN COASTAL DEEP WATER

DEEP SEA WAVE POWER

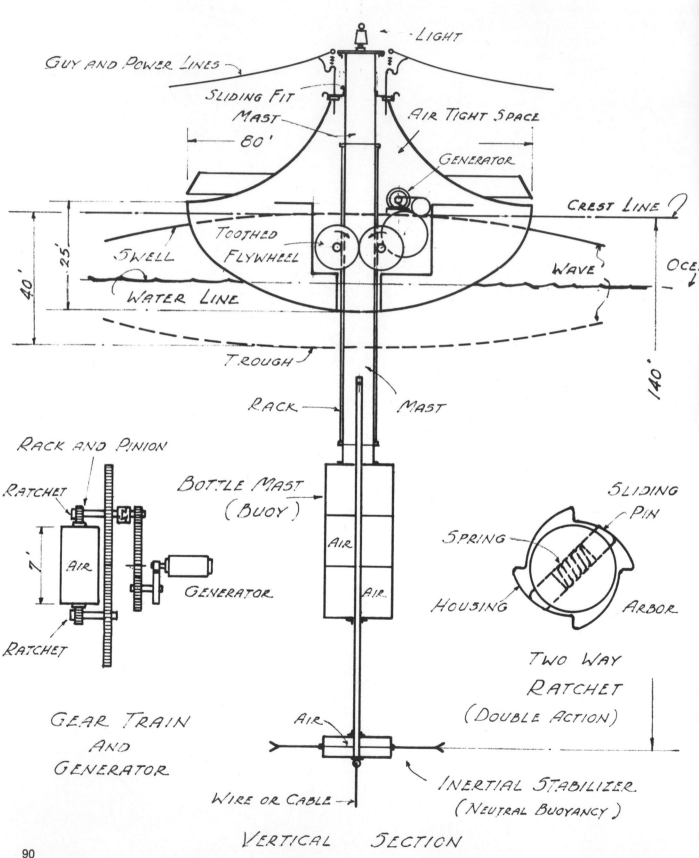

LIGHT

GUY AND POWER LINES

SLIDING FIT
MAST

80'

AIR TIGHT SPACE

GENERATOR

CREST LINE

TOOTHED
FLYWHEEL

SWELL

25'

40'

WATER LINE

WAVE

OCEA

140'

TROUGH

RACK

MAST

RACK AND PINION

RATCHET

7'

AIR

RATCHET

BOTTLE MAST
(BUOY)

AIR

AIR

GENERATOR

GEAR TRAIN
AND
GENERATOR

AIR

WIRE OR CABLE

SPRING

HOUSING

SLIDING
PIN

ARBOR

TWO WAY
RATCHET
(DOUBLE ACTION)

INERTIAL STABILIZER
(NEUTRAL BUOYANCY)

VERTICAL   SECTION

OCEANIC WAVE POWER

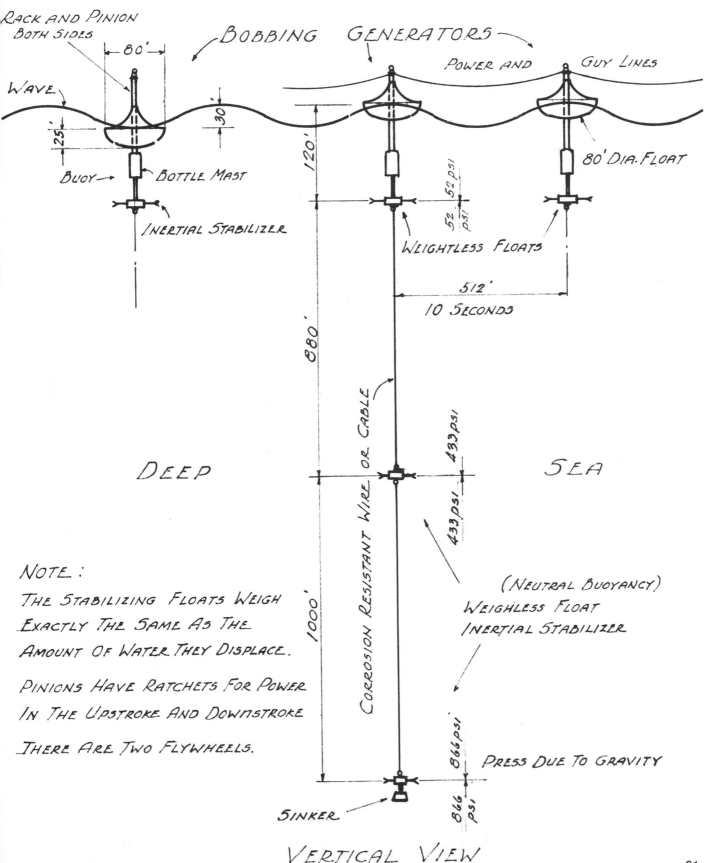

RACK AND PINION BOTH SIDES

BOBBING GENERATORS

POWER AND GUY LINES

WAVE

80'

30'

25'

BUOY

BOTTLE MAST

80' DIA. FLOAT

INERTIAL STABILIZER

WEIGHTLESS FLOATS

52 52 PSI PSI

120'

512'

10 SECONDS

880'

CORROSION RESISTANT WIRE OR CABLE

DEEP

SEA

433 PSI 433 PSI

NOTE:

THE STABILIZING FLOATS WEIGH
EXACTLY THE SAME AS THE
AMOUNT OF WATER THEY DISPLACE.

PINIONS HAVE RATCHETS FOR POWER
IN THE UPSTROKE AND DOWNSTROKE

THERE ARE TWO FLYWHEELS.

1000'

(NEUTRAL BUOYANCY)
WEIGHLESS FLOAT
INERTIAL STABILIZER

866 PSI

PRESS DUE TO GRAVITY

866 PSI

SINKER

VERTICAL VIEW

ELECTRIC POWER FROM THE SEA

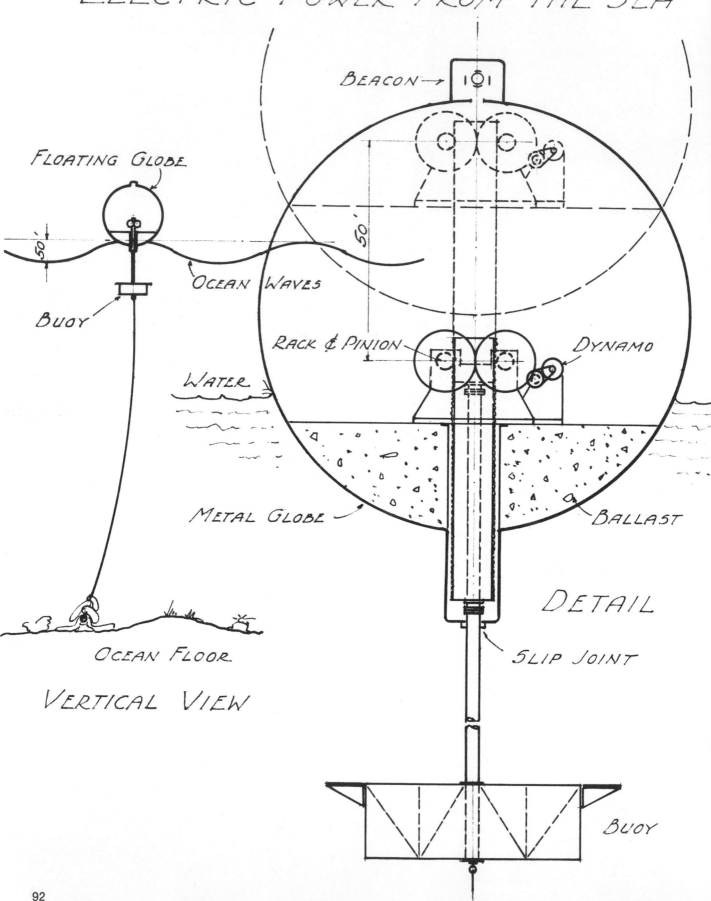

BEACON →

FLOATING GLOBE

50'

OCEAN WAVES

50'

BUOY

RACK & PINION

DYNAMO

WATER

METAL GLOBE

BALLAST

DETAIL

OCEAN FLOOR

SLIP JOINT

VERTICAL VIEW

BUOY

POWER FROM WAVES IN DEEP WATER

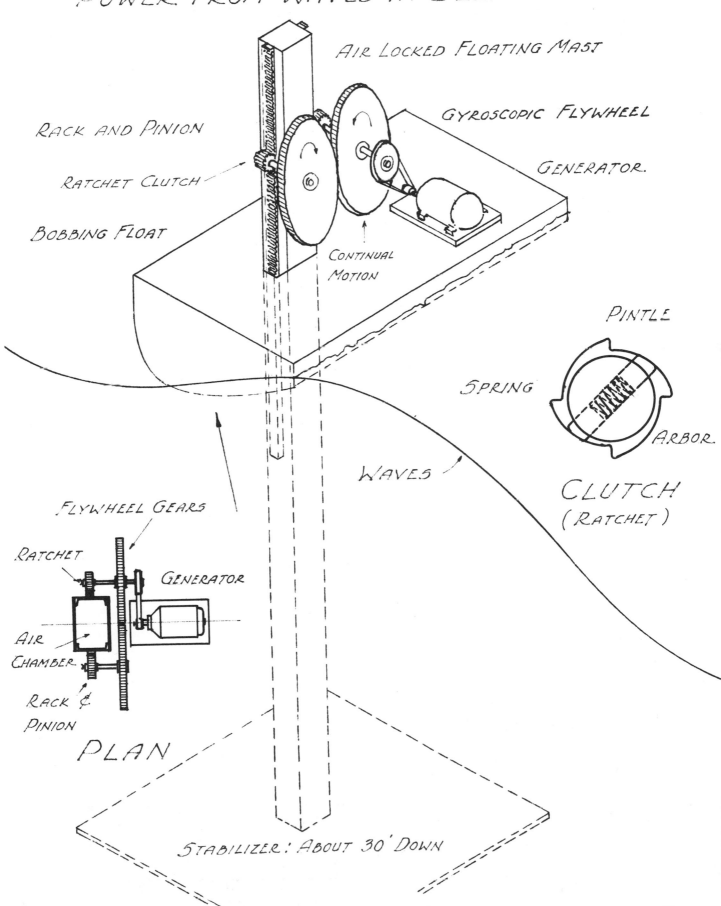

AIR LOCKED FLOATING MAST

GYROSCOPIC FLYWHEEL

GENERATOR.

RACK AND PINION

RATCHET CLUTCH

BOBBING FLOAT

CONTINUAL MOTION

PINTLE

SPRING

ARBOR

CLUTCH (RATCHET)

WAVES

FLYWHEEL GEARS

RATCHET

GENERATOR

AIR CHAMBER

RACK & PINION

PLAN

STABILIZER: ABOUT 30' DOWN

POSEIDON'S WRATH

One of the vital requirements for extracting power from the ocean was the design of a float that would withstand the battering of heavy seas. The ultimate design—called Poseidon's Wrath—was a circular float with an internal waterwheel carrying cups somewhat like those of the old anemometer, with the motion always flowing in one direction. This waterwheel was the basis for converting sea power to electric energy.

POSEIDON'S WRATH

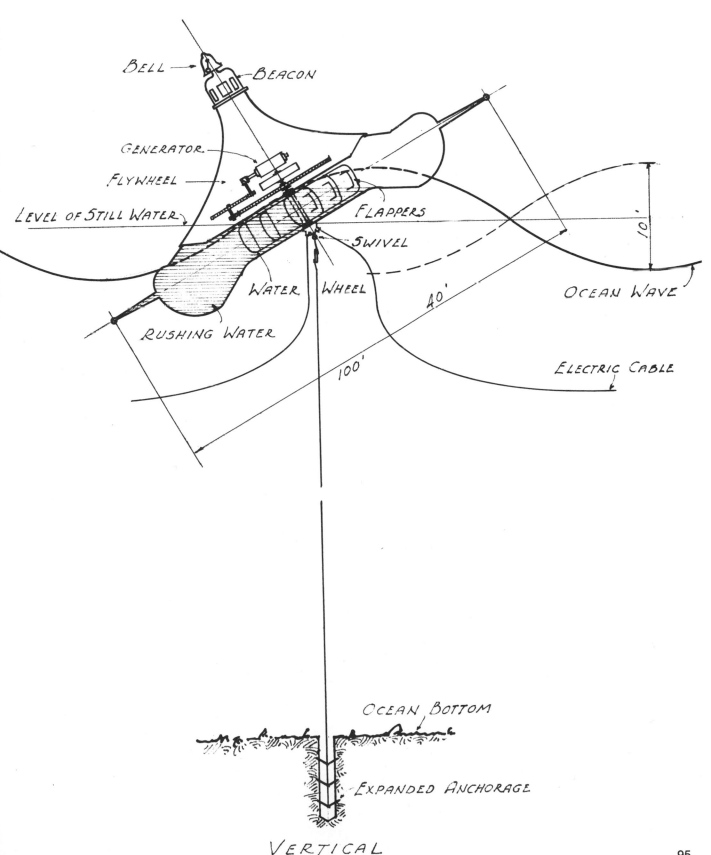

BELL → ─BEACON

GENERATOR.

FLYWHEEL

LEVEL OF STILL WATER

FLAPPERS

SWIVEL

WATER WHEEL

RUSHING WATER

10'

OCEAN WAVE

40'

100'

ELECTRIC CABLE

OCEAN BOTTOM

EXPANDED ANCHORAGE

VERTICAL

THE NEW WONDER OF THE WORLD

From all over the world tourists come to see the new "floating city" of Oceania off St. Raphael in the Mediterranean. You may not have visited Oceania, but you have, of course, seen many spectacular holographs of it.

Oceania is a magic city on the surface of the sea, built as a tourist attraction by the worldwide Disney Enterprises. But there is more than just recreation in the idea of the floating city. The idea for such a city, and the means to build it were first proposed more than eighty years ago.

There is great power in the up-and-down and slightly tilting motion of ocean waves. To utilize that power, requires a transfer mechanism attached to something stable or immovable that serves as an anvil; something on the order of a sea wall or a heavily-laden ship with a deep keel.

Take for example a floating township in the shape of an equilateral triangle, six miles long on each side, having a keel around the perimeter reaching one hundred feet into the water, and a height eight to ten

stories above the sea. Setting over the continental shelf, in water less than one hundred fathoms deep, and away from ocean currents, so large a structure should remain pretty well in place after careful positioning and drill-hole anchoring.

Such a structure would have many uses. It could be served by, and provide a haven for, ships, planes, helicopters, and submarines. It could also usher in the benefits of a marine life culture and industry, as envisioned by Jacques Cousteau and others interested in marine science.

Power would be generated by absorbing the upward force of the waves near the surface, and converting it first to mechanical energy, then to electric energy. This can be accomplished by means of hinged buoyant paddles, ratchets and pawls, a jack shaft with flywheel, and an electric dynamo or generator. The paddles can also be of the platform type, operating with rack and pinion. For continuous power, reliance would be on storage batteries, and possibly on solar heat, with auxiliary conventional power if necessary.

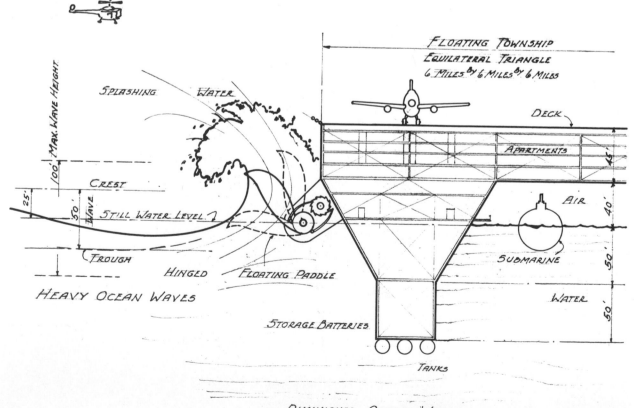

OPTIMUM WAVE POWER FLOATING TOWNSHIP

ELECTRO-PNEUMATIC OCEAN WAVE POWER

Anyone who has ventured into choppy coastal waters in a rowboat or any other small craft knows the power and thumping effect in the boat, particularly when it is pointed athwart the waves. Anything loose will fly through the air, sometimes with smashing force. This violent force is quite different from other kinds of power available from the sea, which are generally strong, but slow in action. Hence we must use different principles to harness it. For instance, the conventional train of gears for changing the power from low speed to high speed or vice versa, can be replaced with compressed air. This is made possible by engaging the principle of the hydraulic ram, the bicycle pump, the hydraulic jack, and others that attain a higher pressure per square inch through the additional effectiveness of impact.

An anchored craft or scow, used for producing electric power in this manner should be of a peculiar design to maximize the force of impact. This suggests a radial keel, so as to accentuate the fore-to-aft rocking effect, loose iron balls or rollers to crash against the plunger of an air pump or a pneumatic ram, and a weighted ball on a mast to induce a snap or whip action, which will prolong and strengthen the forces of impact—the more powerful and sudden the ram action, the higher the pressure of air in the tank. This action will tend to carry over during the wave periods, ranging from five to ten seconds, and will assist in generating power by direct expansion of compressed air.

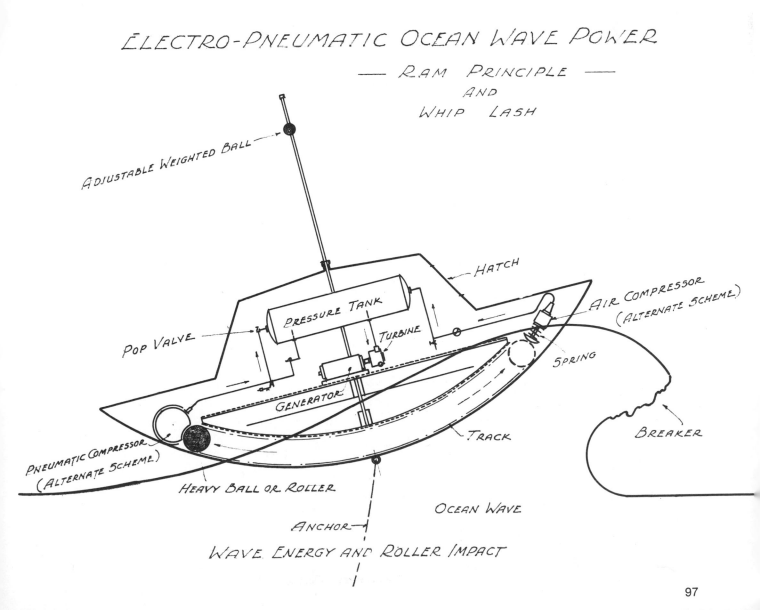

ELECTRO-PNEUMATIC OCEAN WAVE POWER

— RAM PRINCIPLE —
AND
WHIP LASH

ADJUSTABLE WEIGHTED BALL

HATCH

PRESSURE TANK

AIR COMPRESSOR (ALTERNATE SCHEME)

POP VALVE

TURBINE

SPRING

GENERATOR

PNEUMATIC COMPRESSOR (ALTERNATE SCHEME)

TRACK

BREAKER

HEAVY BALL OR ROLLER

OCEAN WAVE

ANCHOR

WAVE ENERGY AND ROLLER IMPACT

DRILL-HOLE ANCHOR

Anchoring an ocean city is only one use for the drill-hole anchor. These strong and firm anchors will have a place in marine agriculture, and in the breeding and growing of oysters, lobsters, crabs, shrimp, clams, and other edible seafood.

This kind of anchoring may also produce power from the waves. Wave action in deep water is generally almost vertical. There is the upward force during the cresting of the wave and a downward force due to gravitation. The upward force offers the greater opportunity, under the present know how, for conversion into useful power. The harnessing of the upward force requires anchorage of some kind. This suggests a heavy weight on a cable or a structure resting on the bottom.

Experience with off-shore oil wells has taught us that

drilling into rock under water is no problem. We know how to create cavities in rock and to reshape metals with explosives. Together, these technical advances enable us to make a sound and solid connection with the ocean floor, and one which will not drift regardless of the weather or the vicissitudes of underwater currents.

The drill-hole anchor can be of several designs and of a proper size and spacing to suit any particular purpose or installation. The cable should have sufficient slack so that in time of a heavy sea, the pull will not be plumb and the stress will not be excessive. The preferred materials for this type of anchorage are: stainless steel, silicon bronze, phosphor bronze, or Monel metal. In some parts of the rig improved plastics may be suitable, particularly for the cable.

DRILL-HOLE ANCHOR

OCEAN WATER

HINGED PRONGS DETAIL   SECTIONAL ELEVATION   EXPANSIBLE CUPS DETAIL

HYDRAULIC PRESS WITH RECIPROCAL BUOYANCY

The qualities of moving water can be made to generate power within a system which involves natural forces working together. For example, one plan calls for a watertight piston in a vertical water tank. The piston is forced down by a priming charge of water. The pressure forces the water below the piston through the risers outside the tank and through the power-generating turbines. The water is discharged on top of the piston.

When the piston reaches the bottom a flapper valve on its side opens, allowing the piston to float upward to the starting level, from which point the cycle is repeated. The operation is analogous to that of the old bicycle pump.

The priming water comes from a reservoir, and is released when the piston has risen to the "ready" position. Since there is no generation of power while the piston rises to the top of the tank, this method employs two or more tanks in series, timed for continuous service.

A more or less similar system can be used in localities where the tides are unusually high. The pressure is provided by the rising and falling of the tide. In this case there is power on both the upstroke and downstroke, but there is a considerable lull at the full flood and ebb stages.

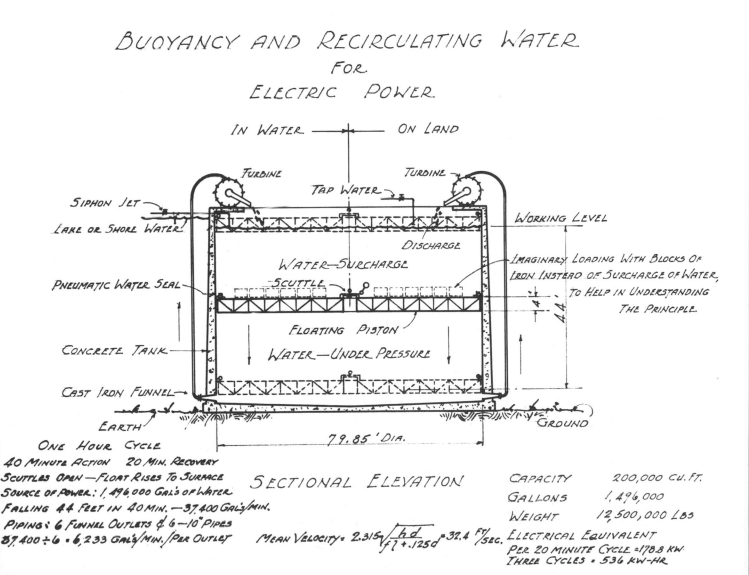

BUOYANCY AND RECIRCULATING WATER
FOR
ELECTRIC POWER

IN WATER ──────── ON LAND

TURBINE   TAP WATER   TURBINE

SIPHON JET
LAKE OR SHORE WATER   WORKING LEVEL

DISCHARGE

WATER—SURCHARGE   IMAGINARY LOADING WITH BLOCKS OF
SCUTTLE   IRON INSTEAD OF SURCHARGE OF WATER,
PNEUMATIC WATER SEAL   TO HELP IN UNDERSTANDING
                                                        THE PRINCIPLE

44'

FLOATING PISTON
CONCRETE TANK   WATER—UNDER PRESSURE

CAST IRON FUNNEL

EARTH   GROUND

79.85' DIA.

ONE HOUR CYCLE
40 MINUTE ACTION   20 MIN. RECOVERY
SCUTTLES OPEN—FLOAT RISES TO SURFACE   SECTIONAL ELEVATION   CAPACITY   200,000 CU.FT.
SOURCE OF POWER: 1,496,000 GAL'S OF WATER   GALLONS   1,496,000
FALLING 44 FEET IN 40 MIN.—37,400 GAL'S/MIN.   WEIGHT   12,500,000 LBS
PIPING: 6 FUNNEL OUTLETS & 6—10" PIPES   ELECTRICAL EQUIVALENT
37,400 ÷ 6 = 6,233 GAL'S/MIN./PER OUTLET   MEAN VELOCITY= $2.315\sqrt{\frac{hd}{f\ell + .125d}}$ =32.4 FT/SEC.   PER 20 MINUTE CYCLE =178.8 KW.
                                                                      THREE CYCLES = 536 KW-HR

MID OCEAN HYDROGEN SEPARATION AND POWER

The shortage of oil set thinking minds to work to find a suitable alternative energy source. The most likely possibilities were atomic fission from such materials as uranium and atomic fusion. The latter, involved the re-uniting of various elements, most probably oxygen and hydrogen which are abundant, of course, in water.

The separation of the hydrogen and the oxygen in water is done with electricity by a process known as electrolysis. As indicated on the drawing, it was learned that the power of the winds and waves could be used to generate sufficient electricity to make the separation. The electrolysis could be undertaken on platforms in the ocean, and tanks of hydrogen shipped to land where it would be fused with oxygen to produce energy at the point of use. It was even learned that favorable locations of the platforms made it possible to utilize solar energy and to add it to the energy generated by the wind and waves.

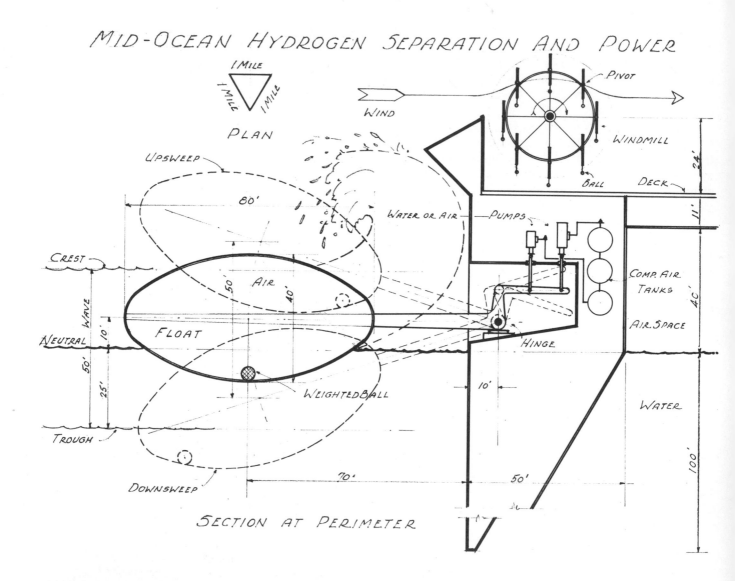

SECTION AT PERIMETER

DEEP-SEA PRESSURE MOULDING

The static energy or pressure of the deep sea can be useful in mechanics. The pressure is great and the energy required to be lowered to and raised from the region of high pressure is relatively small in amount, provided the travel is slow.

Deep underwater, one might still employ the principals of the hydro-press and of rubber forming, in which pressures are more or less equal in all directions. The action is smooth flowing, more nearly like hydraulics than that of the sharp and shattering impact of a power press.

In the past the moulding and joining has been done by putting the work pieces in a rubber bag and placing the bag in a tank where, with high pressure oil, water, or steam, and even air, the work would be moulded or shaped as intended. Good results were achieved due to the flexibility of the rubber bag and an absolutely even pressure exerted in all directions.

The scheme illustrated by drawing is intended as a substitute method for doing the same work, but without burning fuel. The work piece is placed in a soft and pliable bag and lowered in the water with a sinker of neutral buoyancy. The sinker consists of a metal cylinder with an air chamber, suspended with a stainless steel wire on a reel. The weight will sink slowly with gravity, and can be raised at a similar rate of speed if the volume of air in the sinker is in correct proportion. Of course, the slower the lift, the less energy will be needed to raise the work from the high pressure area to the top.

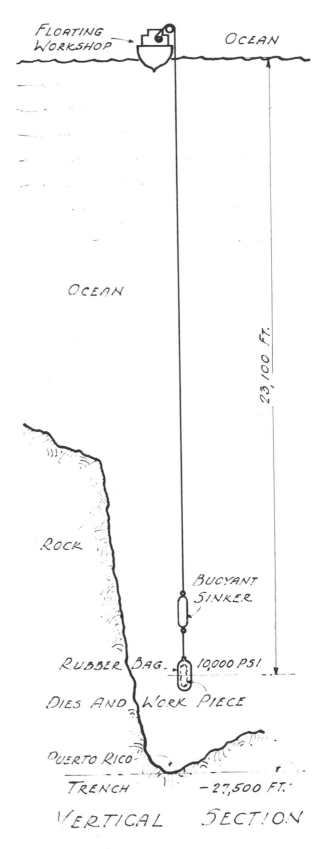

DEEP SEA PRESSURE MOULDING

FLOATING WORKSHOP

OCEAN

OCEAN

23,100 FT.

ROCK

BUOYANT SINKER

RUBBER BAG   10,000 PSI

DIES AND WORK PIECE

PUERTO RICO

TRENCH   -27,500 FT.

VERTICAL SECTION

WAVE POWER AND OCEANIC MINING

We have all seen on holovision, the large floating globes that make up the surface components of oceanic mining units along the continental shelves of the world (the largest ones being those east of Nova Scotia and south of Sakhalin.) Eighty years ago such projects were almost undreamed-of. Here is how a science writer described the possibilities in 1971:

The time is right for study and preparation for the exploitation of nature's riches which abound in and beneath the coastal waters of the world. The continents edges extending into the water and known as the continental shelf, offer industrial prosperity for the future as advanced techniques and equipment become available, and when underwater husbandry becomes urgent and profitable. The continental shelf ranges from a narrow band to hundreds of miles wide, with an average width of around thirty miles. Much of the depth is from fifty to one hundred feet. At present the mining and farming activity is limited to extracting minerals much as magnesium and manganese, and of course the drilling for oil, and the harvesting of shellfish and useful plant life.

Scientific research and development in connection with the design of submersible ships and underwater exploration, are markedly advancing the time when living and working on the continental shelf will be practicable and feasible. At the present stage, a huge diving bell would be suitable for scuba divers in depths of thirty to forty-five feet. For greater depths a thick shell able withstand pressure, and a proper mixture of hydrogenated air is needed to avoid the "bends" or aeroembolism—the bubbling of nitrogen in the blood when a person moves from deep water to the surface too quickly. And to facilitate moving from the ocean surface to the continental shelf, a hermetic system of elevators and submarine structures will be needed.

WAVE POWER AND OCEANIC MINING

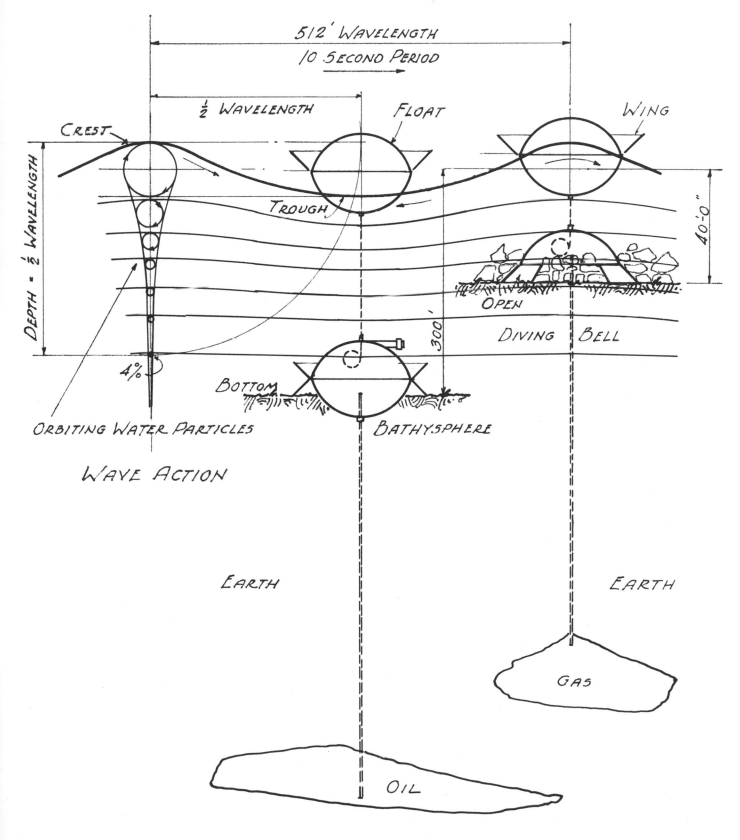

512' WAVELENGTH
10 SECOND PERIOD

½ WAVELENGTH

FLOAT

WING

CREST

40'-0"

DEPTH = ½ WAVELENGTH

TROUGH

OPEN

DIVING BELL

4%

300'

ORBITING WATER PARTICLES

BOTTOM

BATHYSPHERE

WAVE ACTION

EARTH

EARTH

OIL

GAS

GEOTHERMAL EXPLORATION

The possibilities of energy from the depths of the solid earth are still being explored. Deep mining operations conducted during the last decades of the 20th century gave ample indication of the heat of rock formations at various levels. For example, a gold mine in Johannesburg, with a shaft more than 3000 meters deep, showed that rock temperatures ranged up to 59°C.

There was continuing debate about what the earth is made of. Increasingly, thinkers were led toward the conclusion that the center of the earth is semi-liquid, consisting of iron and nickel, with heavier metals like gold, silver and platinum at the core.

As we approach the midpoint of the 21st century we know that exploitation of power from the sun, the wind and the water have moved ahead more rapidly than exploration of possibilities from inside the earth. The depths of the globe constitute our next power frontier.

GRAVITATIONAL & GEOTHERMAL ENERGY

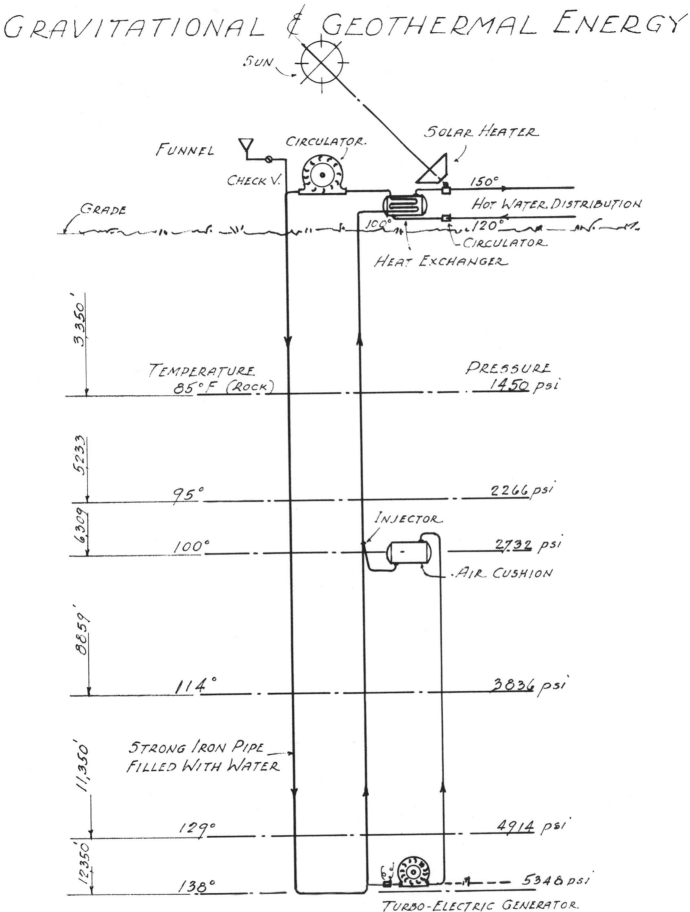

SUN

FUNNEL   CIRCULATOR.   SOLAR HEATER

CHECK V.   150°

GRADE   HOT WATER DISTRIBUTION

100°   120°
CIRCULATOR
HEAT EXCHANGER

3,350'

TEMPERATURE   PRESSURE
85°F (ROCK)   1450 psi

5233

95°   2266 psi

6309

INJECTOR   2732 psi

100°   AIR CUSHION

8859'

114°   3836 psi

STRONG IRON PIPE
FILLED WITH WATER

11,350'

129°   4914 psi

12,350'

138°   5348 psi

TURBO-ELECTRIC GENERATOR.
OR
ELECTRO-CHEMICAL HYDROGEN SEPARATOR

GEOTHERMAL EXPLORATION

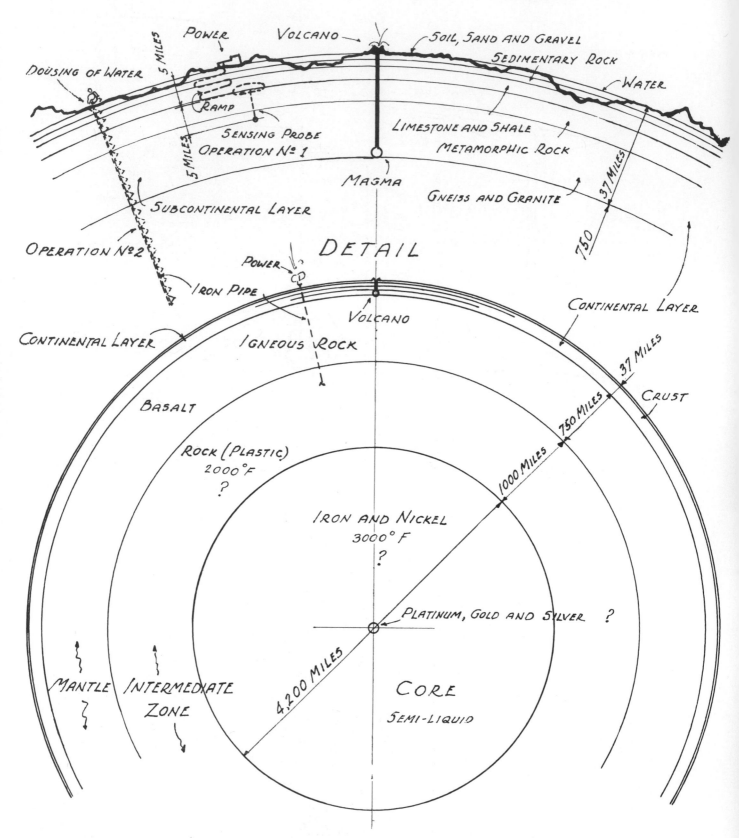

POWER

VOLCANO

SOIL, SAND AND GRAVEL

SEDIMENTARY ROCK

DOUSING OF WATER

5 MILES

WATER

RAMP

SENSING PROBE
OPERATION Nº 1

LIMESTONE AND SHALE

METAMORPHIC ROCK

5 MILES

MAGMA

37 MILES

SUBCONTINENTAL LAYER

GNEISS AND GRANITE

750

OPERATION Nº2

DETAIL

POWER

CONTINENTAL LAYER

IRON PIPE

VOLCANO

CONTINENTAL LAYER

37 MILES

CONTINENTAL LAYER

IGNEOUS ROCK

CRUST

BASALT

750 MILES

ROCK (PLASTIC)
2000°F
?

1000 MILES

IRON AND NICKEL
3000° F
?

PLATINUM, GOLD AND SILVER   ?

4,200 MILES

CORE
SEMI-LIQUID

MANTLE INTERMEDIATE
ZONE

SECTION THRU EARTH

106

HEAT FROM GEYSERS AND HOT SPRINGS

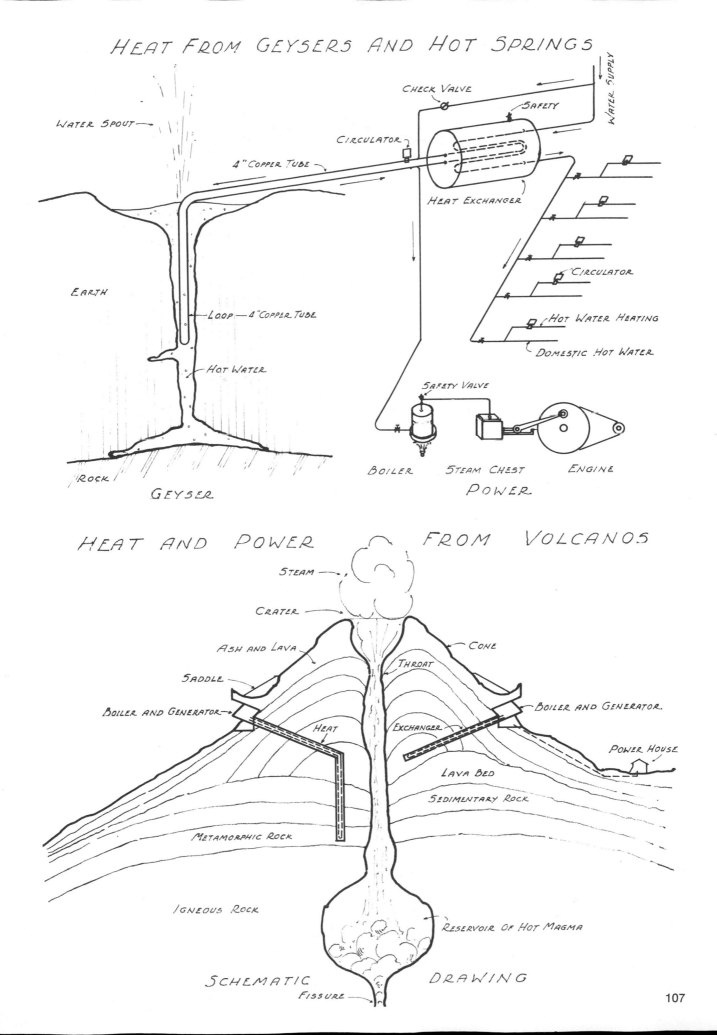

WATER SPOUT →

CHECK VALVE

CIRCULATOR

SAFETY

WATER SUPPLY

4" COPPER TUBE

HEAT EXCHANGER

EARTH

LOOP — 4" COPPER TUBE

HOT WATER

CIRCULATOR

HOT WATER HEATING

DOMESTIC HOT WATER

SAFETY VALVE

BOILER   STEAM CHEST   ENGINE

ROCK

GEYSER   POWER

HEAT AND POWER   FROM   VOLCANOS

STEAM →

CRATER

ASH AND LAVA

CONE

SADDLE

THROAT

BOILER AND GENERATOR →

HEAT   EXCHANGER   BOILER AND GENERATOR

POWER HOUSE

LAVA BED

SEDIMENTARY ROCK

METAMORPHIC ROCK

IGNEOUS ROCK

RESERVOIR OF HOT MAGMA

SCHEMATIC   DRAWING

FISSURE

Solutions to Transportation Problems

Moving of people and goods in the last century became a source of unbearable inconvenience and waste. New and better means of transportation were absolutely necessary everywhere, but particularly in the United States. Here the railroads, once the backbone of the transportation system, had been allowed to disintegrate. Observers began to suggest ways to revitalize this deteriorating resource for the 21st century:

A railroad car should travel at a speed upwards of one hundred miles per hour. An airfoil wing-type design would reduce air resistance to a minimum, and an arrangement of baffles or paved ground could impart a certain amount of uplift so as to reduce the load on the bearings. A wide wheel-span, much greater than the present 4 foot-8½ inch standard gage, would improve the stability on curves and on steep downward grades at high speed. Without special regard for aerodynamics, a railroad car shaped like a large pipe would be the easiest to build and has many advantages. The construction could be light and strong. It could absorb shock without buckling, and would be the least affected by wind and other external pressures. The passenger seats could be mounted on cradles for easy and comfortable riding, and the wheels could be large to reduce friction, and with deep flanges to improve safety.

HIGH-SPEED RAILWAY CAR

There is no mechanical work done with greater efficiency and economy than that of transporting material by means of wheels traveling on rails. A train, like an arrow, splits the air at high speed with a minimum of resistance from head pressure, though its length creates eddying and suction which have a somewhat retarding affect. Since speed and conservation of energy is the order of the day, it seems that new knowledge in aerodynamics should be drawn upon for improvement and maximum efficiency.

The design of the cars and trains will be influenced by progress in the related arts and sciences. Aesthetically there can never be anything more beautiful than a string of cars drawn by a steam locomotive, as in the early twentieth century, rounding a curve at dusk or dawn. Diesel and electric trains have since held sway, but the appearance of these types of locomotive have been anything but attractive.

It is time for an entirely new concept in the design and operation of entire railway systems. A speed of one hundred miles an hour should be normal; the rails should be on easily maintained structures, unaffected by flood, frost, or snow, and above the reach of people or animals; and there should be no "same-level" highway or railroad crossings. The rails should be of wide gage, banked and curved for stability and for absorbing thermal expansion and contraction.

A railroad car with a speed of one hundred miles per hour must be designed to be structurally and aerodynamically correct, light in weight, with correspondingly efficient bearings, and with built-in safety for passengers. A tubular or cylindrical form seems to satisfy all requirements and is the easiest and cheapest to build under our present technology. It requires only a string of offset rolls, adjustable to variable radii, as are commonly used in the fabricating of welded tanks. This concept involves the skin stress principle for structural strength and for lateral stiffness in all directions due to the nature of the cylindrical form. There is also absorptive impact resistance by yielding and folding at the leading edge in time of impact or collision.

Designing for a speed of one hundred or more miles per hour, for cross winds of seventy-five miles per hour, and for the effect of impact, demand that the structure of the components must be strong. The rails should be curved or open-jointed for expansion and the wheels should be relatively large so as not to overheat the journals.

The design makes use of a stubby column which is anchored and aligned in a massive foundation of concrete. The wheels, of which there are four, are large and revolve 105 times per mile, 10,500 per hour, or 175 revolutions per minute. The rails are supported on a triangular open-web truss, which has good vertical and lateral stiffness and strength. The truss also provides transverse shelf angles for added stiffness, and the convenient off-the-ground carrying of pipes and conduits for telephone and telegraph cables, potable water, power lines, and liquid fuels. The gage or spread of the rails, and the double decker cars are proportioned to make the present railroad rights-of-way suitable and feasible.

In exposed reaches where there is the likelihood of up-draughts or treacherous winds, the guide members on top of the cars should be of hook fashion to prevent derailment.

HIGH SPEED RAILWAY CAR
PASSENGER, FREIGHT AND POSTAL SERVICE
AERODYNAMIC AIRFOIL

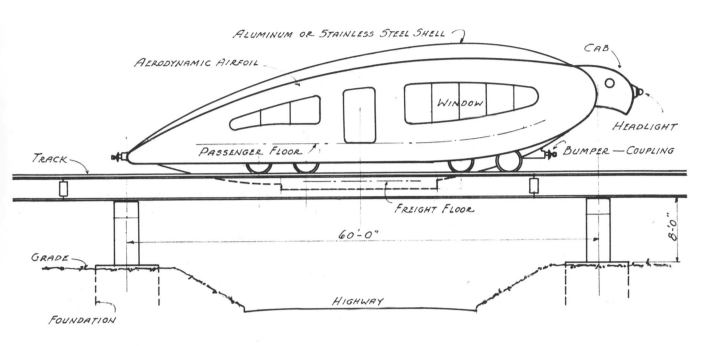

ALUMINUM OR STAINLESS STEEL SHELL

AERODYNAMIC AIRFOIL

CAB

WINDOW

HEADLIGHT

PASSENGER FLOOR

BUMPER — COUPLING

TRACK

FREIGHT FLOOR

8'-0"

60'-0"

GRADE

HIGHWAY

FOUNDATION

SIDE VIEW

DELTA SPEEDWAY — 100 MPH
TRANSPORTATION AND PIPE LINES

COLUMNS

400'

ONE MILE RADIUS

ONE MILE RADIUS

2800'

PLAN OF RAILWAY

20'

60'

DOUBLEDECKER PASSENGER CAR

FLOOR

TRIANGULAR TRUSS

LIGHTS

48" COLUMN

80'

FLOOR

PLATFORM LEVEL

CONCRETE FOUNDATION

RAIL

RAIL

80'

SCHEMATIC ISOMETRIC DRAWING

DELTA SPEEDWAY

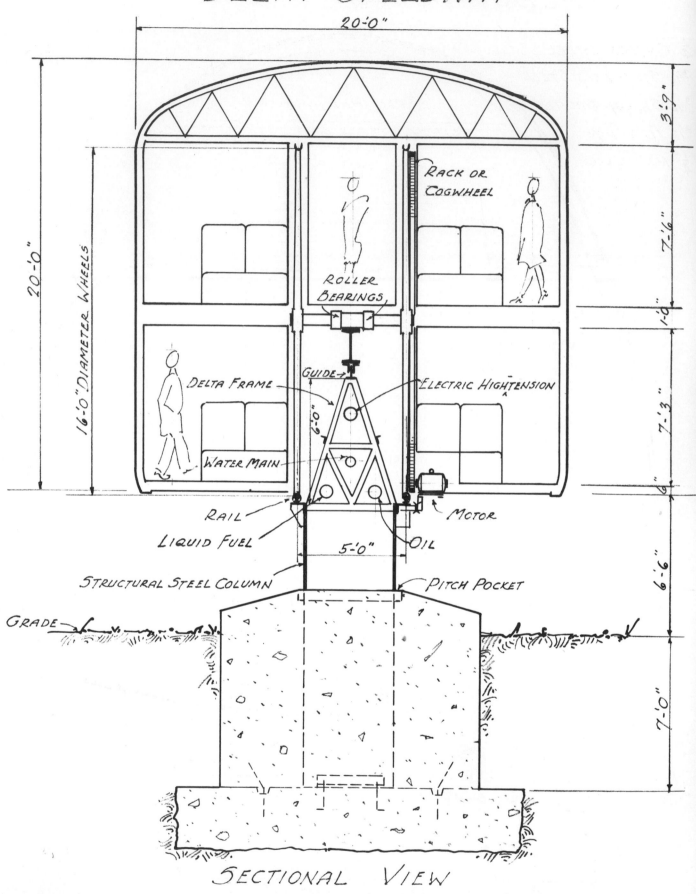

20'-0"

3'-9"

7'-6"

1'-0"

7'-3"

6"

6'-6"

7'-0"

20'-0"

16'-0" DIAMETER WHEELS

RACK OR COGWHEEL

ROLLER BEARINGS

DELTA FRAME

GUIDE→

ELECTRIC HIGHTENSION

9'-0"

WATER MAIN

RAIL

LIQUID FUEL

MOTOR

OIL

5'-0"

STRUCTURAL STEEL COLUMN

PITCH POCKET

GRADE

SECTIONAL VIEW

TUBULAR RAILROAD CAR

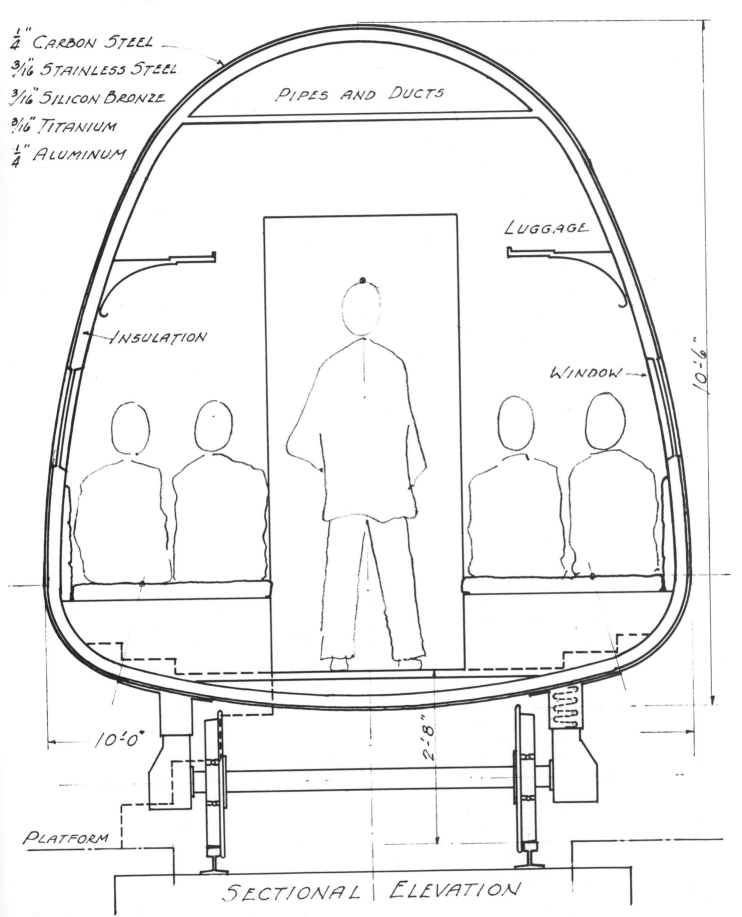

$\frac{1}{4}$" CARBON STEEL

$\frac{3}{16}$" STAINLESS STEEL

$\frac{3}{16}$" SILICON BRONZE

$\frac{3}{16}$" TITANIUM

$\frac{1}{4}$" ALUMINUM

PIPES AND DUCTS

LUGGAGE

INSULATION

WINDOW

10'-6"

10'-0"

2'-8"

PLATFORM

SECTIONAL ELEVATION

BACK TO BASICS IN THE AIR

The first powered flight was achieved more than 140 years ago. In the beginning, man in flight copied the gliding and soaring of the birds. The aircraft was a powered glider. But as jet propulsion was developed, wing size—and gliding angle—were reduced to practical non-existance. The flying mechanism became a self-guided missile. It remained aloft only through the efforts of its power plant. Man was no longer emulating the style of the gull, but rather that of the hummingbird.

This kind of flight was fast, but it required progressively greater expenditure of fuel, with increasing pollution of the atmosphere. There was also greater danger—total reliance on power for flight reduced the pilot's margin for error. And yet jet planes were used for all air travel, over long and short distances.

Around the beginning of the 1970's men began to take a second look at the original models for flight:

For flights of a thousand miles or less jet planes have a slender claim to supremacy. Their advantage is in high altitude flying, and that, is out of place for the short run.

Perhaps there is something to be gained by mechanizing the glider. It would introduce an element of reduced risk during take-off and landing, and it would also require less fuel and minimize pollution. By cruising at levels of a few thousand feet it could serve for inter-city transportation, as well as for medium-distance flying. Smaller airports would be feasible, and would enable this kind of transportation to compete successfully with rail travel, so long as the railroad companies still have to own and to maintain their rights-of-way.

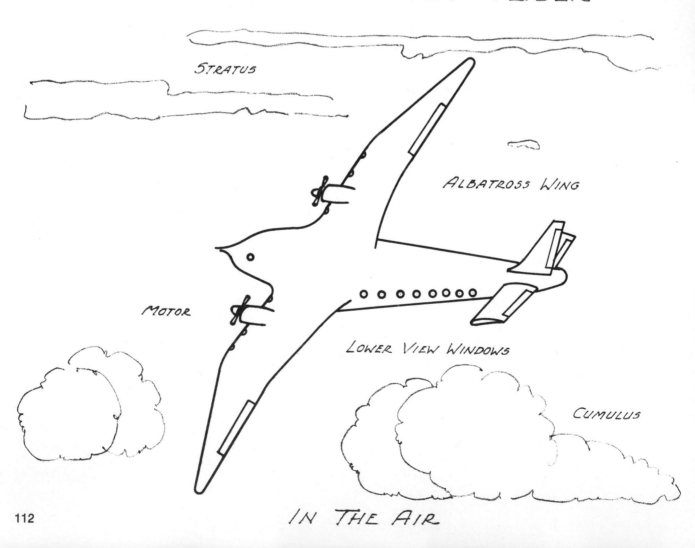

PASSENGER POWER GLIDER

STRATUS

ALBATROSS WING

MOTOR

LOWER VIEW WINDOWS

CUMULUS

IN THE AIR

ENERGY-SAVING AUTOMOBILES

The automobile emerges as one of the principal energy wasters of the 20th century. It was too heavy for the work it did; it burned oil which could have been diverted to more important uses; it polluted the air; and it took many lives.

In this area, as in others, the answer is a return to simpler, more basic thinking. For long-range travel, improved mass transportation was emphasized, with a consequent de-emphasis on the building of superhighways. For short-range use, a small, light-weight car was developed. The declining size of families permitted reduction of seating space. Electricity from storage batteries came back into fashion as a source of automotive power. At first the smaller cars were equipped with auxiliary gasoline engines. As the capacity and efficiency of storage batteries were developed, this auxiliary equipment became unnecessary.

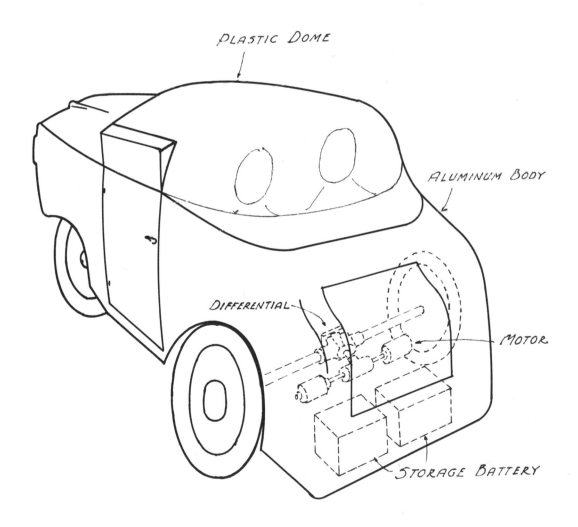

THE THREE WHEELER.

PLASTIC DOME

ALUMINUM BODY

DIFFERENTIAL

MOTOR

STORAGE BATTERY

WEIGHT — 1000 LBS   SPEED — 20 MPH

ELECTRIC AUTOMOBILE

POWER FROM THE FAMILIAR INCANDESCENT LAMP

As a final summary, there is no better example of the creativity that saved our planet than the new use that has been given to the familiar "electric light bulb."

The bulb, or more correctly incandescent lamp, was of course developed by Thomas A. Edison as a means of producing light. But, as the light energy was generated through the application of electricity to a tungsten filament, considerable heat was also produced. This heat was wasted—until it was put to use.

The new application lies in the development of a spherical or global design with a heat lamp in the middle and conical collector-boiler elements on the entire exterior surface of the globe. The clearance on the inside provides a partial vacuum to avoid excessive air pressure and tarnishing of the bright metal of the cones. These elements, arranged in series and cut in automatically when the load increases, have given us a new and highly efficient source of radiant heat.

STEPPED-UP RADIANT HEAT FROM INCANDESCENT LAMP

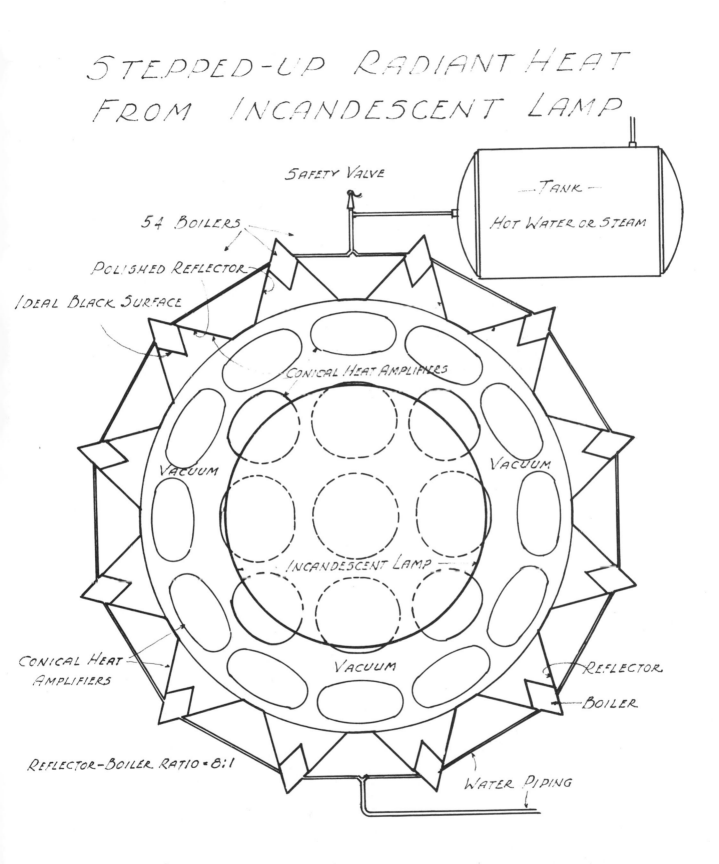

SAFETY VALVE

— TANK —
HOT WATER OR STEAM

54 BOILERS

POLISHED REFLECTOR

IDEAL BLACK SURFACE

CONICAL HEAT AMPLIFIERS

VACUUM

VACUUM

INCANDESCENT LAMP

VACUUM

CONICAL HEAT AMPLIFIERS

REFLECTOR

BOILER

REFLECTOR-BOILER RATIO = 8:1

WATER PIPING

SCHEMATIC DRAWING
GLOBULAR FORM